Advances in 21st Century Human Settlements

Series editor

Bharat Dahiya, Bangkok, Thailand

More information about this series at http://www.springer.com/series/13196

Limin Hee

Constructing Singapore Public Space

Springer

Limin Hee
Centre for Liveable Cities
Singapore
Singapore

ISSN 2198-2546 ISSN 2198-2554 (electronic)
Advances in 21st Century Human Settlements
ISBN 978-981-10-2385-9 ISBN 978-981-10-2387-3 (eBook)
DOI 10.1007/978-981-10-2387-3

Library of Congress Control Number: 2016949630

Printed on acid-free paper

This Springer imprint is published by Springer Nature
The registered company is Springer Nature Singapore Pte Ltd.
The registered company address is: 152 Beach Road, #22-06/08 Gateway East, Singapore 189721, Singapore

Foreword

Some years ago, I was taken aback by discussion about redevelopment of China Square, a venerable area of Singapore, with authorities from the nation's Urban Redevelopment Authority. According to them, the very idea of a square, plaza or similar spatial device was immediately dismissed as being inappropriate. Why? I asked, naively. The area is called a 'square' after all. From this more than momentary befuddlement on my part, it became clear that real understanding of public space in Singapore, despite appearances often to the contrary mostly from colonial days, would require work and interpretative practices outside the way I, as a westerner, normally regarded urban circumstances. This occasion also made me reflect upon similar experiences with public space in Tokyo which had also remained largely incomprehensible to me until I read Augustin Berque's brilliant account of fundamental differences in spatial appreciation in the Japanese ecumen from those in the West.

What the author, Limin Hee, does in this volume is to admirably perform a similar service for the more subtly differentiated case of Singapore. To be sure the Singaporean experience is unique, particularly through its emergence as an independent meritocratic state in the post-colonial era from otherwise difficult and often appalling conditions, alongside the inchoate development of democratic citizens, despite early deep racial divisions. So too, of course, are many other places in the world for different reasons, arguing for equal and equally attentive treatment of their public spaces and beyond mere 'production of space', accounts of somehow 'finished space', and references to familiar yet often moribund 'archetypes' of space.

In defining another way of looking at this world, this book also aims to provide readers with a profound understanding of public space in Singapore and from that, to suggest manners of formulating and reformulating innovative possibilities for future public spaces. The approach, though, is crucial, well substantiated and, again, very welcome. In it, notions of physical and metaphorical space, always present, are maintained in the same order or level of priority. This is important as too often, discussion of public space drifts in one direction or the other and ultimately impoverishes both. Public space here is also seen to be 'constructed' and not simply a matter of 'production' emanating from economic, social and technological conditions and dynamics. Moreover, construction also coincides with the present–future state of being 'in the making', as the author puts it, and not solely an outcome. Overlaps of what are sometimes referred to as 'aesthetic' and 'ethical' considerations are maintained wherever and whenever possible.

Most important, though, all the case materials about places within Singapore, like Orchard Road, Little India, and the Housing and Development Board estates, involve interpretation through what the author defines as 'ground-up construction'. Contrary to assumptions about relatively fixed categories that might come from the social sciences or areas of aesthetic judgment, this form of construction engages with a wide assemblage of material from a wide variety of sources. It embraces, for instance, maps, photographs, written accounts, newspaper articles, interviews, blogs and so on. Furthermore, use of this material is not directed towards trying to simplify explanation and discernment of current or underlying trends. Rather, it is directed, as it should be, towards complicating or thickly layering up interpretation with nuance, subtle reference, surprising moments and uncommon acuity. Finally, at the end of the day, so to speak, Singapore's public realm comes through by way of quotidian reference, day-to-day experience, and the give and take of citizenry on the way to their version of a good life. This applies whether it is the skate boarders on Orchard Road or the storekeepers on Kampong Kapor Road in Little India.

New York
June 2016

Peter G. Rowe
Raymond Garbe Professor of Architecture
and Urban Design, Harvard University
Distinguished Service Professor

Acknowledgments

This book would not have been possible without the support of Mr. Khoo Teng Chye from the Centre for Liveable Cities, Singapore, who understood the value of putting this book out in print, and granted me help in the form of research assistants who interned with the Centre. I would like to express my most sincere gratitude for his kind encouragement and his continued belief in the project.

I cannot express enough thanks also to Prof. Peter Rowe, from the Graduate School of Design at Harvard University, for being my mentor and advisor for many years, and his unwavering support for my work, which was based on my original doctoral thesis at Harvard. I am grateful for his continued collaborations on research in Singapore, even though we are halfway across the planet.

My thanks also to the two talented research interns at the Centre, who worked meticulously to help bring take the book to its final form—Subhas Nair and Wong May Ee. Your hard work and inputs have been greatly appreciated.

Finally, deepest gratitude to Thomas Schroepfer, my wonderful husband, for giving me the fortitude to get the work published, and inspiration by example in having published his own books.

June 2016 Limin Hee

Contents

List of Figures

Chapter 1
Introduction to Study

1.1 Contextualizing Singapore, a Developmental City-state

Singapore is a relatively young city-state, having attained self-rule from the British in 1959, and then independence in 1965. While the modernization process in Singapore can be traced to when the British first colonized Singapore as a trading outpost in the early 19th century, the post-independence economic growth of Singapore has defined the city-state within the global arena.

At the point of self-rule, the nascent state had two primary concerns: political instability and urban deterioration (Garner 1972). While a growing population faced housing shortages, unemployment, and racial tension, the new government also had to handle a rising Communist faction, which was active in the Malayan countryside. In the aftermath of colonial rule, certain policies that the British had employed to establish order (King 1985), as well as existing social and functional features, had to be reconstituted, retained, or revised by the new government of Singapore (Perry et al. 1997).

This included a shift from socializing the segregated racial landscape ascribed according to colonial racial attributes to a more integrated "multiracial" society that had to facilitate the peaceful coexistence of various races, languages, and religions (ibid.). Racial sensitivity was a delicate issue since Singapore's breakaway from the Federation of Malaysia, as the racial composition of the new nation state had a Chinese majority as opposed to the Malay majority of Malaysia.[1] At the time of exit from the Federation of Malaya, Singapore had no economic or political security as an independent nation.

After 50 years of development—marked by rapid economic growth—Singapore had a maturing economy but its achievements had, for the most part, been state-led. Although Singapore could be described as largely middle-class, the government

[1]Singapore was once merged with Malaysia from 1963–1965.

L. Hee, *Constructing Singapore Public Space*, Advances in 21st Century Human Settlements, DOI 10.1007/978-981-10-2387-3_1

warned of imminent social polarization: "Meritocracy underpins the entire Singapore system. But equal opportunities may generate unequal outcomes. As our society matures…these inequalities will become more marked" (Lee 1998).

The government's ability to keep tight control of Singaporean society, which largely accepts its direction and leadership, had been central to Singapore's success (Perry et al. 1997, p. 7). Singapore may be described as the quintessential development state, if we take Castells' definition: "(I)t establishes as its principle of legitimacy its ability to promote and sustain development, understanding by development the combination of already high rates of economic growth and structural change in the productive system, both domestically and in its relationship to the international economy" (Castells 1992, p. 56). On these grounds, nationhood in Singapore can plausibly be found in the shared goal of economic growth.

With the maturing of Singapore's economy and the privatization of wealth, a large "national" middle-class has emerged, many housed in generous "five-room" flats or public "condominiums". They form the nation's most voracious consumers of goods and services, are employed in well-paid white-collar jobs, are Internet-savvy, but remain largely depoliticized, with many of their daily needs—including education, health-care, housing, and retirement—taken care of by the state through policies and structures. In other words, by the early 2000s, a large number of citizens fell into a national middle-class, of which every aspect of life had been wholly constructed by the state. The city-state exemplifies a model of a hegemonic state supported by "weak" publics.

1.1.1 Singapore's Urban Landscape and Public Space

The blueprint for Singapore's urban form—including its public spaces—has been structured around the colonial Raffles Town Plan (or the Jackson Plan) of 1822. Since then, the rapid modernization and urbanization of Singapore in the last four decades had seen unprecedented changes to the urban landscape, and to the form of and framework for the creation of public spaces.

A strategy of decentralization and sub-urbanization served the goal of urban restructuring and redevelopment of the city center for commercial needs. In a 1963 report by a United Nations team of consultants, Charles Abrams, Otto Koenigsberger, and Susumu Kobe, recommended a ring-like development of nodal self-contained settlements with pan-island connection via highways and a monorail. Urban renewal consisting of rehabilitation, conservation, and rebuilding of the city center was vital, with the report noting, "it is evident that urban renewal cannot cure the malaise of overcrowding unless it is made possible even for low income earners to live further afield and reach the Central Business District cheaply and quickly" (Garner 1972, p. 142).

The urban development of Singapore has been manifested spatially by carefully zoned developments such as the Central Business District (CBD), Jurong Industrial Estate, the commercial corridors of Orchard Road and North Bridge Road, the Civic and Cultural District, and Special Planning Areas which are guided by Development Guide Plans drawn up by the planning authority, the Urban Redevelopment Authority (URA). Consequently, nearly 50 % of the island is built up space. Plans for the new Downtown, largely to be constructed on 372 ha of reclaimed land to create 6 million m^2 of new commercial space, would meet Singapore's demand for the next 50 years for key financial and business offices to replace the aging CBD (URA 2000).

The ethnic '*kampongs*' or enclaves that Raffles had put in place in his Town Plan have generally undergone gentrification—in Chinatown, Little India, and Kampong Glam. Increasingly, retail spaces in these conservation districts have been taken up by boutique businesses and trendy offices, replacing traditional businesses. Lush business and science parks have also sprung up on many parts of the island, changing the landscape of industrialization in Singapore. The mass rapid transit system already links the new towns around the island to the Central Area, while a system of land-based highways creates efficient aerial road linkages island-wide. The physical form of the city has become more fragmented and complex, and variety has become an aesthetic of the new landscape.

Large-scale implementation of modernist housing paradigms was introduced by the British and adopted in the 1950s and 60s—and the experimental nature of such an extensive urban venture was without doubt unprecedented in Singapore. It is also significant that the continuities of the historical traditions of urban public spaces were disrupted with these new modernist paradigms of mass housing. There had been no attempt to provide any response to either the traditional housing environments, or to the use of outdoor spaces. This break was sudden and effected on a large scale: large segments of the population in colonial Singapore were displaced and relocated to high-rise public housing to open up land for redevelopment in the city centers.

In Singapore, public spaces take on multiple identities as the result of specific interactions and articulations of contemporary socio-urban processes that come together in time. As such, these spaces do not display a coherent identity but are contingent with everyday practices at global and local scales, the intersection of these being at the scale of the city. My discussion of public spaces in Singapore is thus situated in this shifting and dynamic context.

Since self-rule, the issue of public space planning in Singapore has been political. There was a conscious, unwritten rule that large spaces ought not be appropriable for demonstrations, race riots, or mass protests that were phenomena rampant in the inchoate era of self-government. The nature of government in Singapore, the pragmatic economic emphasis on many fronts—including national development, and the adoption of neo-Confucian ethics in the forging of national values—have created a particular social milieu. This social milieu is reified, albeit not directly, in the forms of public spaces created within the last few decades. As noted, there had been a clear disjuncture in the forms of public spaces in the city—first through colonial intervention; then a period of post-colonial rapid modernization of the city extending to the social-engineering of public spaces in the new towns around the city-state; and finally the contemporary imaging of new public spaces in a global city-state.

Another aspect of this study is the examination of the design of new public spaces in high-density, high-rise environments, and to evaluate if these spaces have enabled some form of continuity in the social uses of space. Nowhere else has the Congrès Internationale d'Architecture Moderne (CIAM) model of functionalist mass housing been adopted on such a large scale, and with apparent success in housing so many. Yet, few studies have investigated the forms of spaces in between housing and their coded embodiment of social agendas and spatial practices that have emerged in these spaces. The design of housing units and prototypes has been the subject of many studies by architects and designers, and social scientists have studied the political, social, and economic impacts of the public housing phenomenon. However, there has been no study regarding how spatial form and structuring in these housing contexts relate to spatial practice and deeper cultural orders.

Sentiments of rising nostalgia in the 1980s and 90s coincided with the perceived "arrival" of the nation in terms of material accumulation, but somehow lacking in "spirit and soul" (Kong and Yeoh 2003, p. 132). Restoring identity to local culture, the revival of collective memories, and the re-enchantment of place are often the thrusts of urban restructuring in the building of the new symbolic cultures of cities. In Singapore, the development of the city center has come full circle. While in the 1950s and 60s, efforts were concentrated on hollowing out the city for its redevelopment as the economic engine of the fledgling city-state, planners are now attempting to bring back vibrant places in the city.

New spaces for consumption, such as the 'post-traditional' space of Orchard Road, herald not only new social spaces in the city, but also new archipelagos of public space. The new centrality of cities, as prophesized by scholars of global cities (Borja and Castells 1997; Sassen 1998), need to be pre-empted by making cities more civic, along with the development of thickly-laid networks of mobility. The new importance accorded to public spaces and urban aesthetics again foregrounds the meaning of cities as sites of meeting and creating civic sense and identity, as well as of imaging the new centrality of cities, i.e. for its symbolic value in the promotion and expression of citizenship.

The understanding of public space necessitates the a priori understanding of the national culture, and the questioning of its existence. If a Singapore culture exists, does it exist in public space, and would we recognize it as such when we see it? Perhaps it would be easier to assess this culture in contrast to others rather than as a thing by itself. For example, a Singaporean culture versus a Japanese culture, or a French culture, or American culture of public space. The state attempts to direct national cultures, and, in turn, various forms of mass media and entertainment tend to reflect the state. Public space is the site for the expression and re-production of a Singaporean national culture, embodying physical manifestations and representations of such a culture, then re-producing these in forms of spatial practice. It is often in the reflection of national culture that constructions of public space occur.

The urban development of Singapore and the conditions of its urban and cultural landscapes incite broad questions about its public space:

- Have the new spaces in the city and in the new towns enabled extant spatial practices to exist? Have these practices adapted to new spaces? Or have entirely new and emergent spatial practices resulted from new imaginative typologies—if these exist at all? In other words, can we connect the imagination of spaces to the imagination of practices that inhabit the space?
- Just as we understand that new forms of social networks and associational life have emerged, what are the structures of association extant in relation to the forms of public spaces in the city and in the living environment?
- Are relations between the state and the people of Singapore played out in public spaces? What are the relations among the different entities that make up Singapore society? If these are manifest in public space, what are the dynamics of these relationships?
- How do the scales of public spaces, their form and design, and their functional and iconographic programs relate to the experience of everyday life? How do they avail themselves as transcendental public spaces of urban life?

In order to investigate these aspects of public space in Singapore, it is necessary to frame the research not just within the contextual development of public space here, but also to situate it within the current discourse on public space.

1.2 Framing the Study

The use of the term 'public space' juxtaposes the term 'public' and its shifting definitions with notions of 'space' that presents by itself a whole plethora of definitions, depending on disciplinary discourse. The terms for the design of good public spaces are often elusive and vague, and the translation of the abstract ideals and ideas of the public realm into real physical space is fraught with difficulties. However, design has an important role in the making of places that ultimately enables the creation of vibrant public life in the city and within urban communities.

As such, well-considered spaces are no longer part of the abstract urban flux but become places that enable the establishment of spatial practices constituting a larger cultural order replete with meaningful associations and narratives—knowing what it means to inhabit the city. However, the design of good public spaces has to come out of a framework of understanding that is not limited to the physical but situated within epistemic discourses, historical specificities, and social referents. As such, any useful research will have to draw from theoretical perspectives and discourses on public space, as well as lessons from extant urban and spatial practices.

While notions of rapid modernization and globalization tend to take on meta-narratives in explaining urban processes, this study attempts to understand overlapping phenomena of political, social, and cultural origins, within the framework of the design and creation of public spaces. I will both describe the qualitative and operational forms of these spaces as well as their social referents and frames in the Singaporean context, and situate this discussion within the larger framework of the discourse on public space.

1.2.1 Objectives

The discursive values of constructing meanings through the interactions of experience, physical form and spatial practice are germane to the study. The dynamic interactions of processes, program and practice and the resultant spaces shaped through the contingent historical specificities are constructed within forms of contextualization of public space. As such, the pertinent research objectives are:

1. To develop a conceptual frame of reference to construct the discourse on Singapore public space
2. To construct a preliminary model of Singapore public space through analyzing case studies
3. To locate the modes, methods of production and representation of these public spaces within the rapidly changing urban context
4. To situate these constructions of public space and its possible trajectories within the larger discourse on public space, and to examine the viability of such a construction of public space.

1.2.2 Dimensions of Analyses

Space, as Foucault suggests, is no longer to be treated as "the dead, the fixed, the undialectical, the immobile" (1977). Although the design of spaces in no way directly establishes spatial practices, space is a powerful enabler of social processes; there is a congruence of spatial forms with deep underlying cultural orders. However, culture (as defined as transmittable social practice) is plastic and malleable, and thus adaptable to new spatial environments, which are morphologically different from traditional ones.

Familiar models of public space include those predicated on the relationship between the form of public space and the use and socio-cultural meaning of these spaces in the development of typologies. Squares, boulevards, public gardens, and arcades not only made the city readable, but held meanings and uses that were understood by everyone. However, such a relationship of typology and meaning has not been part of Singapore's urban history, a city that has seen a more checkered type of development than a continuation and persistence of spatial types. Here, the examination of the shifting meanings and use of places over time, the deformations of typologies of spaces over time, as well as the importation of new typologies and their reconstitution is more relevant to the analysis of Singaporean public space than a mapping out of a history and taxonomy of types of public spaces.

Specifically, the study examines the various scales of organization and engagement of selected spaces within the context of Singapore to understand the dynamic relationship between formal and social structures. Through case studies of programmatically distinct public spaces in Singapore, the following will be explored:

- Processes (historical events, political, cultural) that shape public spaces—though such linkages are often indirect and not temporally parallel with the creation of public space. Public spaces have also changed through time, taking on new roles or new meanings, or else moving from being a space for everyday life to one of symbolic use, as a spectacle and image condenser. Sometimes overt control strategies in public spaces have resulted in sterile, uninteresting spaces. But life between buildings in turn finds new tactical territories for expression, in the continuing contestations of control and use of public spaces.
- *Parti* or formal and spatial qualities of public space, with their constituent design elements, shape, enclosure and scale, are often embedded within ideas of how space should appear, its social roles, how communities should interact, and distinctions between public and private realms. The types and range of public spaces and their interface with state and private interests also inform us of ideological concerns—being often the manifestations of urban design politics.
- Program—functional, aesthetic, iconographic or symbolic programs embedded within spaces—and how these translate (or otherwise) as meaningful experience to the public who inhabit such spaces, and how the design of spaces enhance or impede such programs. Public spaces have often been infused with social objectives, especially at the level of community public spaces in new towns. How these have changed with shifting social objectives reflect interesting parallel experimentation with manipulating space to institute social practice.
- Practice—the associated spatial practice which is shaped through the quotidian or spectacular use of public spaces at both individual and collective levels, as well as the larger cultural orders which are forged through such spatial practices over time. An important question to ask would be how public spaces frame civic experiences, and have the potential to become the associational spaces of collective actions or social transformations. The experience of public space in the housing environment on an everyday basis gives these spaces the potential to form part of the social life of communities.

While process, *parti*, program and practice are outlined here as the initial and abstract categories for thinking about public spaces, the methodology for the research by no means dictates that the material fits neatly into these epistemic definitions. More likely, the material would overlap, fall in between categories or would be difficult to categorize, since the phenomena constituting public space are necessarily complex.

There is moreover a temporal layering of experiential space that permeates the analytical lenses outlined above; these are the spaces between here and there, the gray zones of the urban realm. Such "non-places" have been definitively put into the realm of experiential space through the work of anthropologist Augé (1995). Augé argues that many spaces that we pass through lack the meaning of classical anthropological place, where meaning is derived through human activities. Instead, the phenomenon of over-individualization[2] affects the use of public space, which is seen less as social spaces than areas, which people use individually. Such spaces abound in the sphere of mobility, and examples of these in the context of this research would include transit spaces, temporary places, five-foot ways, back lanes of shophouses, void-decks and similar informal, interstitial spaces.

The research framework is premised on a thick synthesis of the analytical dimensions to construct notions of public spaces within the historically, politically, and culturally contingent context of Singapore, and to relate the role of design as an enabler (or otherwise) of spatial practice. Such an understanding will form the basis for formulating design trajectories for good public spaces within the political and social milieu of this study, and in also shaping relevant issues for the design of new public spaces. The capacity of public space to form part of the everyday life of people, and in enabling new forms of associational life in the city and in the housing environment forms the discursive platform for the research.

1.3 Definitions and Ways of Thinking About Public Space

In the ensuing discussion on public space, a broad definition of what public space entails has been adopted—which does not submit to distinctions between physical space and metaphysical space, but maintains them within the same epistemological plane. The concept of public space adopted in this research is one that is 'constructed'—not only as a "product" of political, social, cultural and urban agencies through historical processes, but also as the locus of individual and collective experience. In other words, public space is understood as an epistemic device, as material object, as mimetic structure, and as imagined construct. I find this latitude useful as a way for thinking about the complex concept, as well as for the reasons of subjective accessibility argued by art historian and critic, Rosalyn Deutsch:

> For once we pose the question of the right to the city democratically, we dispense with appeals to grounds of meaning that exist outside of discursive intervention and instead acknowledge that our responses to the question depend on our representations of public space. For this reason, I do not distinguish between so- called discursive or "metaphorical" space and so-called concrete or "material" space.

[2]Over-individualization is one of the three conditions of *Supermodernity*, which Augé describes that I find relevant in this discussion.

> Those who insist on such a distinction often called the second space "real", the first "unreal", and then accuse anyone concerned with subjectivity in representation of abandoning real political struggles taking place in real public spaces (2000, p. 81).

The realms of discussion within this research also move from the political, social, cultural, urban and to design, in the belief that these are all somehow 'spatialized', for according to urban geographer, Edward Soja:

> While it may be easy to grasp the idea that everything spatial is simultaneously, even problematically, social, it is much more difficult to comprehend the reverse relation, that what is described as social is always at the same time intrinsically spatial. This inherent, contingent and completely constituted spatiality of social life (and of history) must be persistently and explicitly stressed, lest it be forgotten or submerged (2000, p. 8).

Public space tends to be viewed as built environment or as physical containers of human activity, shaped by design professionals as well as by social and historical processes, concentrating these into spatial expressions. To 'spatialize' in this context means to locate physically, historically, or conceptually social relations and practices in space, and not to "produce" material space out of ideological, social, economic or technological means. To restrict the definition to mere "production", argues Soja, disregards the effect of space on notions of social and cultural life, in other words, the "dynamic, generative, developmental and explanatory qualities" of space (ibid., pp. 8–9).

Public space, as the object of investigation in this study, is 'constructed' rather than read as a finished 'product'. The object of study is not so much the production of space from macro-economic perspectives, or through ideological and social struggles, as the processual or 'in-the-making' aspects of public space, and exploring the potentials for transformation within such processes. As such, practices in space, or spatial practice, exist within the concept of public space as simultaneously in the domains of imagination and action. By understanding public space both as material and phenomena, I hope to unearth public space's potential for design, and conversely, design's potential for public space. The objective of the study is not to be precise, but to texture richly the notions of public space, as the concept is highly complex and the methods for which to study it are imperfect, approximate, and often offer many open-ended findings.

1.3.1 Ethics and Aesthetics of Public Space

Current trends in the thinking about public space coalesce along two main discourses—one of ethics and the other of aesthetics. However, this is not to say that these do not overlap or cannot be considered together. Instead, by placing contemporary thoughts on public space along these two perceptible streams of influence, the divergence of discourses in the present time is highlighted. This reflects a change of attitudes about public space from more classical definitions, in which the ethics and aesthetics of public space were part of the same conception, rather than separate discourses.

The archetypes of public space in the history of European cities consisted of spaces such as squares, plazas, boulevards, gardens and the marketplace, all of which conjure up fixed typologies in the imagination of those who inhabit these spaces, as well as fixed meanings, codes of civilities, and associations with these types. These configurations of public spaces were important elements in ordering the city, with their fixed immutable typologies and the legibility they lent to urban form. With the onset of industrialization and modernization, new city planning strategies (e.g. zoning), and the rise of private capital, forms of public spaces became increasingly amorphous, subjected to new definitions, functions, and antagonisms with private interests. There was an overall proliferation of types and forms such that the immutability of form and meaning was changed forever.

Theories of public space abound, from theorists and practitioners alike, but these discourses do not come together as unified theory, nor do they approach offering any distinct typology or definition of ideal public space. For the purposes of this discussion, I would like to consider theory together with the practice of public space, such that theory is neither a priori nor a posteriori to space. Rather, theory acts as an epistemic device for the discussion and understanding of space. From here, what I term as the 'ethical' discourse on public space, often conflated with discourse on the more abstract and de-spatialized public realm, takes a distinct trajectory from the 'aesthetic' discourse, which is more often than not a practical discourse, with architects and designers as their main proponents. While wary of generalizations, the 'ethical' discourses tend often to be associated with discourses of nostalgia, or narratives of loss, while the practical discourse tend to be imbued with a sense of optimism about the reinstatement of public space through architecture and urban design.

1.3.2 The 'Ethics' Discourse

I find it useful to think of the discussions on public space within two allegorical categories of space: the first, in which space is 'performative', a conception of public space as a theatrical 'stage;' the second, in which space is 'filled', a conception of public space as a 'collective space'. Of course these categories are entirely abstract conceptions, but are useful for structuring the discussion.

Public space as 'Stage': The notion of public space as a 'stage' presupposes that space is undefined to start with, and is taken up with informed actors operating in concert—even if pluralistic views are held—through knowledge, communication, and action. An example of the first instance of public space as 'stage', would be the Habermasian public sphere (1989), where actors, whose personal differences are "bracketed", practice rational-critical discourse on political matters. Therefore, public space has the potential to be the space of communication rather than domination, and thus has democratic potential. Identities are, however, not constructed in public, but are created within the private realm. Hannah Arendt's agonistic "space of appearances" (1971), as exemplified by the Greek agora, is one in which moral and political greatness are displayed by the actors, who seek recognition and acclaim in public space. Both Jürgen Habermas and Arendt describe the deterioration of the public realm through the rise of the free market, the break-up of the constituents of the public, and the breakdown of the notion of a common good.

To compensate, Habermas advocates the use of communicative action through organized civil society in debate with the state to bring about transformations, while Arendt advocates "associational public space", in which men act in concert to push a cause, and where freedom can appear, coordinated through speech and persuasion. The new public space in both the Habermasian and Arendtian definitions neither resides nor can be represented by buildings and spaces, but is instead summoned into existence by social actions.

Political theorist Mouffe (1999, 2000) blames the decline of the public sphere on the loss of collective forms of identification, or the specificity of subject positions in which to fully participate in democratic debate. The onus on consensus-based politics, shared values, and the disappearance of antagonism generate new allegiances from forms of identification such as of ethnic, religious or nationalist nature. In other words, political debates become conflated with issues of morality. Mouffe calls for the transformation of antagonism into "agonism", so that vibrant confrontation can be reinstated in the 'stage' of public space.

While the social actions described by Habermas, Arendt, and Mouffe are executed by organized groups through speech, persuasion, and action, theorists like de Certeau (1984) and Bourdieu (1977) discuss the potential for resistance at the level of the individual body. Through the actions of the human body on the 'stage' of everyday life, practices inscribed in space constitute the "tactics" of the weak or dominated against the broad "strategies" of dominant space.

Place, in de Certeau's definition, is an ordered structure provided by the dominant order through which its power to organize and control is exerted. Such an interpretation allows us also to deem place as the embodiment of hegemonic strategies, which impose and define behavior. Thus, places are planned and built to control and organize movement and behavior of subjects in the interests of the dominant. However, within and against these impositions, people construct their spaces by the practice of living in place. Thus, space is practiced place.

De Certeau stresses that political conflict is involved in such confrontations of social interests and is central to the creation of space out of place. In other words, the practice of creating space out of place can be interpreted as tactics—the "art of the weak"—that are indeed practice. The victories resulting from tactics exist in moments of acting out or performance, and often do not produce or accumulate in concrete form anything that can be stored or is permanent. However, the products of tactics should not be seen as less valuable than the products of strategies in the creation of public space. What they manifest is resistance to domination. However, such a resistance does not always mean opposition, for it may also take the form of circumscribing order, the creation of identities within homogenizing place, or creative narratives within the routine and mundane.

Public Space as 'Collective Space': The notion of space as 'collective space' presupposes that space is full to begin with. 'Collective space' is a state in which groups occupy domains and co-exist on a competitive basis with other groups. Interactions between groups may be spontaneous, but differences exist and may at times lead to antagonism. Collective space tends to be more social than political in nature, and may approximate the marketplace more than the agora. The 'experiential' rather than communicative dimension of public space is foregrounded.

An example of this second allegorical category of public space is what Lofland (1973, 1998) terms the "public realm", by which she means a social rather than physical environment that is not always geographically rooted. Lofland's public space is populated by strangers—it is a "world of strangers", for whom space forms the communication medium, structures how communication will occur, and can be altered by the content of the communication (1998, p. 186). Public space, by virtue of the existence of different groups, is a contested realm. The mutual proximity of multiple spheres in public space leads to exchange and confrontation. Spaces dominated by particular groups are considered "parochial" spaces and exist as a realm between the public and private. If these public spaces become a collectivity of parochial spaces, i.e. strangers no longer exist in these realms, then a decline of public space occurs.

In this respect, Setha Low conceives the plaza as a microcosmic public space populated by diverse groups competing for cultural representations:

> Plazas are spatial representations of Latin American society and social hierarchy. Citizens battle over these representations because they are so critical to the definition and survival of civil society. Plazas are also centers of cultural expressions and artistic display reflected in their changing designs and furnishings. And finally, plazas are settings for everyday urban life where daily interactions, economic exchanges, and informal conversations occur, creating a socially meaningful place in the center of the city (2000, pp. 33–34).

Low argues that the aesthetic, political, and social aspects of the (South American) plaza change through larger sociopolitical forces, but are also contested through conflicts over the use, design, and meaning of the space. As a cultural form, the plaza is spatialized culture represented through "multi-vocal" narratives interwoven with experiential and mimetic structures in public space. Rules of social encounter are engaged in the plaza, thus structuring social interactions. The social and cultural forces "produce" the plaza and create "an environment and an ecology" which is also colored by personal experiences and perspectives in space, so that space becomes transformed or socially constructed. Low maintains that the relationship between the social production of space, through historic, macro-sociopolitical, economic, and professional understandings, and the social construction of space through use and meanings, "is dialogic rather than dialectic, in spite of the high degree of conflict and contestation…that the plazas act as containers, thus permitting resistance, counter resistance, and change to occur publicly and with relative safety" (ibid., p. 131).

The actions of competing, or under-represented groups in public space, are also embodied in feminist discourse on public space (e.g. Fraser 1990), as well as in discourse regarding the positions of marginalized or minority groups in space (e.g. Caldeira 2000). Discourses about new cultural landscapes, including work by post-colonial theorist and anthropologist Arjun Appadurai, describe spaces like "ethnoscapes" in the age of globalized media and transmigration, which are essentially "fluid and shifting landscapes of tourists, immigrants, exiles and other moving groups or persons" (1996).

While the two allegorical categories of public space serve as useful devices for framing models of spaces, the two are not mutually exclusive, and may overlap or be situated within the other. One could easily imagine characters like Benjamin's flâneur[3] or de Certeau's bricoleur[4] slipping between the 'stage' and 'collective space' and operating within both discourses. In this sense, when the 'stage' becomes crowded, competing groups may contest for space and identities.

Discourses of Nostalgia: Discourses of nostalgia and narratives of loss abound and often decry the end of public space or the evacuation of the public realm. There are many streams within these discourses, and they stem from different values, perceptions, and meanings attributed to public space, such as the loss of *civitas* (Sennett 1977), legibility (The New Urbanists), authenticity (Sorkin 1992), communication and action (Arendt 1971), democracy (Deutsch 1992), and the rise of consumerism (Zukin 1995). Just as the rise of the free market and private interests were deemed to have divided the concept of the public, the proliferation of spaces associated with consumption led to the perception that the 'public' dimension of spaces and the meanings associated with fixed typologies were 'lost'. The theming of large-scale shopping malls and centers of amusement after classical space typologies, for example as plazas, markets and "main streets", has had an impact on the 'real' city through discourses of "authenticity" (Hajer and Reijndorp 2002).

Public spaces streamlined for consumption by specific groups, such as tourists, or spaces overly "instrumentalized" for specific programs are criticized for the loss of potential as social mixers. Similarly, spaces under extreme surveillance become too "defensible", as private territories become too crippling for social interaction. Mike Davis, in his work, "City of Quartz", comments on the deliberate control of pseudo-public spaces through design and organization: "Ultimately, the aims of contemporary architecture and the police converge most strikingly around the problem of crowd control. As we have seen, the designers of malls and pseudo-public space attack the crowd by homogenizing it. They set up architectural and semiotic barriers to filter the 'undesirables'. They enclose the mass that remains, directing its circulation with behaviorist ferocity, it is lured by visual

---

[3]Walter Benjamin's flâneur, as described in *Das Passagen-werk* (1983) is found on the Parisian boulevards: "there was the pedestrian who wedged himself into the crowd but also the flâneur who demanded elbow room and was unwilling to forego the life of a gentleman of leisure" (p. 54). The flâneur is that character who retains his individuality while all around are losing theirs and derives pleasure from his location within the crowd, but simultaneously regards it with contempt.

[4]A bricoleur is a person, a tinkerer who engages in "making-do", as described by Michel de Certeau in *The Practice of Everyday Life* (1984).

stimuli of all kinds, dulled by muzak, sometimes even scented by invisible aromatizers" (1990, p. 257).

However, many of these arguments can also be used to suggest the factors leading to these developments as forces of democratization exercised through spaces. The rise of consumer groups, and choices available to the public, are democratically additive rather than cause a decrease in "public-ity". The creation of new boundaries and borders, and different types of public spaces may be perceived as new ways of organizing public space, with multiple definitions of 'public'.

The rise of individualization is a widespread phenomenon in many developed societies, and an outcome of affluence and better access to information. People no longer have to be defined by their traditional bonds with institutions such as class, labor and gender roles. Instead, as defined by sociologist Beck (1986, 1992), individualization also means "first the dis-embedding and second, the re-embedding of industrial society's ways of life by new ones in which the individuals must produce, stage, and cobble together their biographies themselves". With an increase in individualism, the effect over the long term on forms of public space remains open to speculation. The discourses of nostalgia in public space may, after all, not really be histories of deterioration, but accounts of the changing forms and roles of public space.

Anthropologist Marc Augé, in describing the phenomenon of the emergence of "non-place", does not take a nostalgic view of lost space, but is intent on exploring the nuances of this new category of space that he has identified. "Non-place" has its own dynamics and characteristics and emerges from the abstract and material development of urban life under conditions of flux. The potential of non-place as a form of public space opens up new trajectories for investigation.

We return to one of the original "nostalgic thinkers" of public space—Hannah Arendt, whose account of public space has been one of *Verfallsgeschichte*, or a history of decline. Feminist Benhabib (1992), in a re-reading of the Arendtian discourse, sees the redeeming of nostalgia through her methodology of "story-telling" as an attempt to think through history in sedimented layers of language. By so doing, she identifies "moments of rupture, displacement and dislocation in history" so that one can "set free the potentials of the past" (ibid., pp. 73–89). In other words, by employing the lens of nostalgia and emotions, the framing of the past enables one to reflect on the present in new ways and to transform it.

1.3.3 The 'Aesthetics' Discourse

Architect-theorist Rem Koolhaas has this "nostalgic" view of public space in the contemporary city:

> I think we are stuck with this idea of the street and the plaza as public domain, but the public domain is radically changing…with television and the media and a whole series of other inventions, you could say that the public domain is lost. But you could also say, that it's now so pervasive it does not need physical articulation any more. I think the truth is

> somewhere in between. But we as architects still look at it in terms of a nostalgic model, and in an incredibly moralistic sense, refuse signs of its being reinvented in other populist or more commercial terms.... you can go to these cities and bemoan the absence of a public realm, but as architects it is better for us to bemoan the utter incompetence of the buildings (1996, p. 45).

However, despite his seemingly resigned tone, the projects from Koolhaas' own practice show an active interest in the architect's role as an interpreter and mediator between private commissions and public interest. They defy the bemoaned loss of the public realm by re-creating it within architectural space—with the Seattle Public Library project, for instance. A new architectural 'stage' is erected where public space is re-defined through a deliberate 'congestion' of programs. The same attitude that public space no longer exists in its own right but is a component of the interaction of architecture and the urban fabric, is again pronounced in the proposal for the Beijing CCTV project, in which two mega-buildings within the city structure social relations, flow, and communication between the city and architecture.[5] Similarly, the works of Jacques Herzog and Pierre de Meuron embrace the concept of public space within the architectural "stage" as a means of integrating architecture with the urban, such as with the Beijing 2006 Olympics Stadium and the De Young Museum in San Francisco.[6]

An exhibition held in New York in the summer of 2003 by the Van Alen Institute titled "*OPEN: New Designs for Public Space*", featured "the most celebrated urban spaces, places that contribute to creative metropolises, combin(ing) large-scale operations and small-scale experiences, major infrastructure with recreation and culture, and grand civic events with intimate daily interaction".[7] Works of internationally renowned architects, urban designers, and landscape architects in international projects were organized around main themes which dealt with access and transparency, the role of information in public space, new functions of streets, new social spaces for events and activities, memory in place, parks, and temporary artistic interventions in space. "Many of the projects do not fit traditional categorizations of public space. Not all of them are open 24 h a day, a number charge an entrance fee, some are only partially open to the public, but these are among the real conditions of public space, especially in the 21st century..." (Gastil and Ryan 2004, p. 9). Said Zoë Ryan, the exhibition curator, "The exhibition shows projects from small interventions in previously underutilized spaces, such as a media center in a subway and an artificial island in a river, to large urban projects like an 11-mi pedestrian and cycle path with gathering spaces and a major cultural center that provides a new meeting ground in the heart of the city. The projects prove that public space is still a critical and thriving aspect of city life and one in

---

[5]The Seattle Public Library is described in " Seattle Public", pp. 138–149, and the Beijing CCTV in "CCTV" pp. 480–509, in Koolhaas, R. (ed.) *Content*, Köln: Taschen, 2004.

[6]Jacques Herzog argued in his lecture at the Graduate School of Design, Harvard University, on March 18, 2005, that the potential of public space has been overlooked in architecture, and consciously uses urban public space as a device for his architecture projects.

[7]Press Release by VAI June 15 2003.

which new approaches to design are fostering interaction and exploration". Criticism leveled at the work of designers often dwell on the abundance of aesthetics, and it is not unexpected that such a remark was made in the press: "'Open'-ness is a nice idea, yet even a show devoted to it demonstrates that contemporary planners' embrace of the public realm remains more theoretical than actual" (Washington Post 2005).

In Singapore, in the same summer of 2003, a major design competition for public spaces around the city was launched—with the objective of "generating awareness for the need for good public spaces" (Skyline 2003)—which drew much public interest in the design and programming of public space. In the same year, the Public Spaces and Urban Waterfront Master Plan for the city center was launched, and the island-wide Parks and Waterbodies Master Plan helped to identify projects which have since been completed, such as the Southern Ridges, Marina Bay Waterfront Promenade, Gardens by the Bay, Woodlands Waterfront and Punggol Waterway. With the increased use of social media platforms, the URA reached out to the public, using crowdsourcing through its 2013 *PubliCity* initiatives, to gather ideas to rejuvenate public spaces:

> Public spaces play a critical role in our plans to create a good quality living environment. They serve as venues for communities to gather and interact, and help to strengthen social identity and foster community bonding. Through this project, we hope to invite the community to celebrate public spaces, and at the same time participate in giving ideas on how to make them better," said Ng Lang, Chief Executive Officer of URA.[8]

The perennial question remains: can good public space be deliberately created? The perceived divide between the aesthetics and ethics of public space is difficult to bridge, and current approaches to the design of public space tend to emphasize aesthetics over more amorphous needs not easily translated into design vocabulary. While not prescribing normative design directions in public space, this study aims to construct notions of existing public space from the ground up so as to have a better understanding of the dynamics and practices in public space. It is with this understanding that ways to approach the problem of the design of public space may be formulated in innovative and novel terms.

1.3.4 Situating Singapore Public Space

Many of the theories and discourses on public space originate in Western Europe or the United States, while sites like Singapore have developed from entirely different histories and exist within different phases of urban development. Although there are, and will be, instances where existing discourses and models of public space overlap imaginatively with public space in Singapore, some of the a priori

[8]This was part of the *PubliCity*: *Your Ideas for Public Spaces competition* that was organized by the URA. See https://www.ura.gov.sg/uol/media-room/news/2013/nov/pr13-76.aspx.

conditions necessary for the existence of public space in the idealized western, liberal democratic model do not exist or are not comparable within the Singaporean context. The historical founding of Singapore, its colonial history and occupation during WWII, the contrived nature of Singapore's existence as an independent city-state, and the inchoate development of democratic citizenry are unique to Singapore's historicity and its urban development.

As such, it is necessary to develop a methodology for this research that does not assume an a priori model of public space, but to construct such a model within the Singaporean context. The discourse on public space as understood in the rational-critical tradition, when applied to Singaporean public space, might yield criticism but would not be adequate in identifying alternative processes that allow for transformative potentials in public space. On the other hand, discourses that assume the competition of defined groups in space may gloss over the shifting or multiple identities of the 'weak' publics in the Singaporean context, instead of focusing on the competitive dynamics in public space. This study will instead situate fragments of discourses and trajectories within the Singaporean context, and under present conditions to re-assemble a discourse for a new construction of public space.

1.4 Towards a Methodology

> ...that space and architecture are at their deepest levels of significance always beyond description and, therefore beyond any epistemological frame whatever. So our alternative is...an alternative to epistemological frames...it follows that knowledge of urban space can only be accessed by engaging with particular architectures and spaces (Bishop 2004, p. 2).

To engage with a culture of public space is to try to pin down something which essentially shifts, changes with time and in fact, occurs in temporal episodes, and often leaves no traces in space itself. But to capture merely the physical state of public space is to examine an empty stage devoid of performance, gestures, meaning, and spirit. The purely aesthetic, symbolic, or formalistic interpretations of public space are not sufficient for understanding public space as it leaves out the "lived" dimensions—spatial practices which have transformative potential in public space. By situating public space within the cultural landscape of everyday life, the notion becomes assembled through the thick texturing of the experiential—that together with memories, emotions, and stories, paint quite a different picture. As explained by Low, the mainly western discourse on public space may also omit "local stories" of place:

> ...macro-political interpretations...are not sufficient for understanding the plaza because they leave out the people who use the plazas and its importance in their everyday lives. These perspectives exclude the indigenous archaeological and ethno-historical past, as well as the memories, stories, and meanings of plaza life. Instead, Eurocentric explanations of the origins of plaza architecture and formalistic readings of political symbolism determine what is known, while the local stories go untold (2000, pp. 33–34).

However, purely ethnographic approaches, from observations and documentation of use also leave out important aspects of public space. As explained by Highmore (2002), conducting "...everyday life studies will require from the start more than one perspective. It also suggests that one cannot look simply from above, by looking at established and dominant assemblages of writings supplemented by empirical ethnography. After all, the kinds of knowledge that can be assembled through observation and questioning are also likely to express that which can be most easily articulated. It might also be necessary to grasp the texture of the everyday through experience in art and literature. Everyday life experience may in fact be situated between the kinds of attention that would focus either on subjective experience or on the institutional frames of cultural life".

1.4.1 A 'Constructed' Notion of Public Space

In order to destabilize established methods of examining public space, such as "post-occupancy" surveys and ethnographic documentation, I propose an approach to construct public space from the ground up, by assembling materials from various sources to suggest how thick, textured "diagrams" may be used to indicate or "diagnose" possibilities of understanding, leading to new assemblages of the notion.[9] This research proposes to 'construct' public space with the intention of experimenting with new techniques of public space. Not only is physical space the object of study—but visual documentations in the form of photographs, maps, and sectional drawings, written materials in the form of books, newspaper articles, interviews, commentaries, pamphlets, popular literature, stories, poems, quotations, personal itineraries, and logs such as documented in Internet blogs, activities in space, including events, parades, festivals, and everyday life routines—are juxtaposed and re-assembled in this study. After the new "cartographies"[10] or "diagrams" of public space are constructed and their familiar meanings given new insights, they are "diagnosed", not through a process of reduction, but through a process of "complex-ification".

The methodology is somewhat similar to what Walter Benjamin did with the drafts of his uncompleted opus magnum, *Das Passagen-werk* (2002), which was to juxtapose a great number of materials, including illustrations, quotations, thoughts, and commentaries about the social and cultural history of nineteenth century Paris: arcades, salesmen, catacombs, iron construction, advertising, the flâneur, prostitution, etc. Benjamin's passion as a collector of information was well-documented (Arendt 1992). Because the new constructions were deliberately dis-synchronous

[9]The terms "diagram" and "diagnose" are explained by John Rajchman in his essay, "A New Pragmatism?" and are used here in the sense of Gilles Deleuze's "abstract machine" that tries to capture the processes and form critical breaks with continuities to re-assemble new configurations of thought.

[10]This is another term attributed to Gilles Deleuze.

and were not lodged in a narrative or discursive structure, they could be re-manipulated to suit the demands of the changing concept of "the present" (Buck-Morss 1991, p. 336). Similarly, the practice of Brechtian theater is one whereby "no single perspective or mode of presentation is ultimately privileged" (Highmore 2002, p. 23), so that plural and contradictory accounts can be assembled. Thick narratives, offered through "techniques of interruption" render the Brechtian world "unfinished and open to exploration" (ibid.).

As such, the method through which I present the materials follows a visual rather than linear logic, in the sense of the montage (Buck-Morss 1991). Information is collected from various sources and different media and assembled within "frames": the narrative images and anecdotal texts within each framing, together with a series of urban diaries with an emphasis on the verb 'to do'—construct the concepts of public space and its itineraries from the ground up. The montages or frames are segues into the discussion of public space, and do not form urban typologies or any fixed categories of space, but bring together diverse ideas and phenomena.

Instead of functioning as a starting point of fixed typologies and practices for public space, these frames act as dynamic montages compressed in time to identify certain themes in spatial practices, supplemented by textual materials. Without reducing the images and textual materials to essentialized forms, the intention is to use these images and texts in a novel way—as concentrated information, compressed in time, framed and juxtaposed to bring out hidden meanings and themes in public space. Here, public space is defined by experiential and temporal dimensions, instead of fixed urban typologies at the same time as the specificities of the urban locations are documented. The concepts of the public and space use are constructed from the ground up—beginning from the level of the body—its postures, positions, clothing, being together or apart, its actions, and so on.

1.4.2 Situating Spatial Practice

The 'construction' of public space on many levels, from the scale of the urban to the scale of individual everyday life routines, re-assembled as process, necessarily foregrounds dimensions that may previously be hidden from the official discourse. Just as public space could be used to perpetuate the dominant discourse and practices in the interests of the state, it is also the site of counter-practices or acts of resistance. The question arising from the methodology in excavating public space through the dismantling of familiar structures is whether identities in space should be preserved or submerged.

While de Certeau's notion of practice (1984, 1998) submerges the "subject" and privileges uses over users, Bourdieu's '*habitus*' (1977) redefines the subject through acts of habitation. Bourdieu's concept of the *habitus* is useful in framing the material culture of everyday life to relate the habitat,[11] (in)habitant, the processes of habitation and habits. The notion of the *habitus* is enriched by the thickness and density of embodied practices, daily life narratives, relationships and experiences, as well as the objects of everyday life. Some cultural theorists have termed such material density "texture"—such that the materially constrained realm of everyday life is compensated for and contradicted by the intensity and density of experiences, practices, and objects that are packed in it. This is concomitant with de Certeau's work on the building of cultural density around objects and practices that evoke what has been described as the cultural imaginary, or imaginative fiction—an individual's account of being in the world and of situated-ness in time and memory. Such a concept of practice or *habitus*, with its dense embodied notions of practice, experiences, and objects, is constitutive of the subject being constructed from bottom up. This stands in opposition to the conventional social science definitions of subject identities and notions of multiple subjects. One of the problems in de Certeau's rendition of the *Practice of Everyday Life* had been the onus on the "poetics that articulates activities rather than express identities—a poetics of 'uses rather than users'" (Highmore 2002, p. 156). This could be understood as the attempt to transcend traditional sociopolitical frames of reference of the subject as constructed from structural differences, such as class, gender, and so on. On the other hand, the practices of everyday life, both within and against social space and social order, are the means to construct from below, the identities of difference of the social actors.

Such a method of constructing identities presents a means that is concretely embedded in everyday life instead of theories of multiple subjectivity categorized through top-down splits based on structures of economy or social politics. Even poststructuralist analyses of nomadic subjects moving across categories are predisposed to create porosities through these categorical boundaries, rather than through focusing on the material practices of individuals within space and time. John Fiske has termed such differences "popular differences"—these being "produced by and for the various formations of people: they oppose and disrupt the organized disciplined individualities produced by the mechanisms of surveillance, examination, and information, which Foucault has shown, are the technologies of the mechanism of power" (1992, p. 161).

These differences are socially and historically contingent, i.e. context specific, and are not the products of biological or psychological differences. They are produced through the culture of everyday life, within the realm of practice and experience—this being not the realm of differences being acted out, but the producer of these differences. If we persist further along the Foucauldian trajectory of social order being imposed through the control of the body and its behavior,[12] it becomes clear that the body is the material site upon which top-down impositions of

[11]A habitat is the social milieu, constituted by both social spaces and the practices of those who inhabit these social spaces.

[12]This is detailed in Michel Foucault's *Discipline and Punish* (1977).

power and bottom-up constructions of identities and differences take place. As such, the body is socially situated, and its behavior is produced in space and time. Practices of habitation extend the body into the habitat and relate it to other similarly or differently habituated bodies.

The concept of the *habitus* thus helps us to understand also how identities can be constructed through practice, and that the 'habitat' is not a social space but a process with historical and social specificity. It thus offers us a framework for situating subjects within concepts of practice. It helps us describe ways of living within social space and therefore public space, through which social identities are formed. Such a theory of practice helps us gain some insight into how the specificities of everyday life and spaces are the sites for the creation of social identities and differences, and how a 'practical' (as derived from practice) concept of public space can be developed through such a construct.

1.4.3 Summary: Framework and Methodology for the Study

With the notion of 'constructed' public space as the object of this study, the "coherent narrative" and "rigorous argument" of social science research may not be the most suitable method or form of presentation of the material.[13] Within a framework of 'constructing' public space, this study attempts to bring together phenomena such as the relationship of the state and its citizens, elements of Singapore society, identities of transient entities in space, changing spatial practices, and urban development. Without privileging a single source of information or mode of presentation—textural, pictorial, or graphical material—a method that allows thick description of public space, approximating the process of montage, is employed in the research. The qualitative and operational forms of these spaces as well as their social referents and frames within the larger context of the state are diagrammed and montaged, but in unfamiliar ways, to make visible processual qualities.

Not particularly engaging the notion of public space as a form of 'production' within a political-economic model, this research does not attempt to break down the boundaries of the political, social, and cultural, but strives to bring these realms into proximity through experience and practice in space. The categories of physical space and metaphysical space are not distinguished within this study. Neither is public space defined as an a priori realm of political order or communicative action, nor does it remain as an abstract notion or representation. Instead, through the conflating of practical conditions and abstract notions, public space is regarded as a multi-dimensional space of discourses and encounters. In considering the

[13]Highmore (2002) also puts forward this argument for approaches that need to grasp across different registers for the study of everyday life.

processual phenomena of public space instead of the finished product, temporality is intrinsically embodied, so that public space resides in time as well as space.

Mostly, public space is situated within urban everyday life, within the realm of actual encounters, imagination, exchanges, and effect. Public space is addressed not at the level of actual content, but at the de Certeau-esque level of form, as modes of practices and sometimes bricolage. The methods used would be imperfect and approximate, and the findings or "diagnosis" open-ended and at times elusive. The study uses actual sites of public spaces as vehicles of processes, chosen to reflect the different configurations of urban public space in Singapore, and the different entities that inhabit these spaces. Tangible and intangible material will be collected, inserted and juxtaposed within the constructed frames of this study. Trajectories of public space, when made, will not actually be of imagined futures, but of an expanded present, re-assembled.

Chapter 2
The State, People and the History of Urban Public Space in Singapore

2.1 Shaping the Singaporean Public

> If Singapore breaks up, it will never come back. It's man-made, it's very contrived to fit the needs of the modern world and it has to be amended all the time as the needs change. The moment it no longer fulfils that role, it will begin to decline. I would put it at one chance in five.
>
> Lee Kuan Yew[1]

As evidenced by Lee's statement, Singapore—the developmental city-state—is an artificial product custom-made to fit in the modern world. The process of designing Singaporean culture and in so doing, shaping the Singaporean public, has been a highly self-conscious one. The history of Singapore's nation building can be described as a series of periods and juxtaposed events, punctuated by "founding moments" (Bishop 2004). The prevailing thread of continuity in its relatively short history of nationhood is the persistent state of crisis that the nation perceives itself to be facing.

Singapore is essentially a small country, a city-state borne out of a traumatic past and facing an uncertain future in a rapidly changing region in the world; occasionally Singapore is sport for a bout of 'Singapore-bashing' from its neighbors in the region, such as China, Taiwan, Malaysia, Indonesia, or even from superpowers such as the USA. However, it is difficult to dismiss Singapore as pastiche, such as the labels that describe Singapore as "Disneyland with a death-penalty" (Gibson 1994, p. 51) or that "the city resembles a clean and efficient theme park" (Branegan 1993, p. 36). At many levels, one can discern the existence of a coherent Singapore society, despite the presence of multiracial tensions underlying the harmonious, multiracial society as reflected in the national rhetoric.

---

[1]This is a quotation by Lee Kuan Yew from Han, Fernandez and Tan's book, *Lee Kuan Yew: The Man and His Ideas* (1998).

L. Hee, *Constructing Singapore Public Space*, Advances in 21st Century Human Settlements, DOI 10.1007/978-981-10-2387-3_2

Although a 'prehistory' of Singapore exists, the history of modern Singapore undoubtedly began with its British founding, for the colonial occupation of Singapore effectively wiped out the traces of its earlier ephemeral empires and rural Idyll.

2.1.1 Occupations and Influx: Colonial Singapore, Immigration, and the Japanese Occupation

Singapore, under the British colonial government from the 1819 to the Japanese Occupation of 1942,[2] was not different from other such colonies where the practice of 'divide and rule' prevailed. Stamford Raffles, then Lieutenant-Governor of Bencoolen and agent of the East India Company, identified the strategic importance of Singapore as a hub sited between the trade routes of the western and eastern worlds. Singapore's urban development under British colonial rule was to facilitate its role as such a shipping hub. However, Raffles first had to deal with the native population and the greater problem of rapid migration from other parts of Asia into the newly founded Singapore in the nineteenth century.

The Jackson Plan for Singapore of 1823 segregated the colonials from the indigenous and other ethnic groups to control and supervise the behavior of these groups. This spatial segregation helped to maintain the power relations between the 'dominant and dominated', and at the same time, keep distinct the existence of the various ethnic groups to help the colonial authorities to anticipate how each group would behave (Hee and Ooi 2003, p. 82). Each immigrant group was designated an area under the new plan: the Chinese were settled west of the Singapore River, with Chinatown further divided into plots for various dialect groups; the *temenggong* (Malay ruler) and his followers were settled west of the commercial district; while affluent Asians and Europeans lived in a residential area adjacent to the government quarter.

The segregation of the races occurred not only in space, but also in time as it was occupied through the vocations of these groups. Many migrant workers, mostly males, hailed from the region in search of a better life, and found jobs as menial workers. Some had come from China,[3] to escape poverty and natural calamities such as floods and droughts; some came from India, the Malay Archipelago and

---

[2]Although Singapore attained self-rule from the British in 1959, the colonial reign was effectively over by the time of the Japanese Occupation.

[3]The largest Chinese dialect group in the late nineteenth century (and now) were the Hokkien who were traditionally involved in trade, shipping, banking and industry. The next largest group, the Teochews, dealt with agricultural production and distribution, as well as the processing of crops such as gambier, pepper, rice and rubber, etc. The Cantonese worked as artisans and laborers. The two smallest groups, the Hakka and the Hainanese were mostly servants, sailors or unskilled laborers. Historical information regarding the occupations of ethnic groups can be obtained from the very comprehensive report of Country Studies: Singapore, Dec 1989.

Indonesia to work on plantations; at the same time, the British sent convicts from India to provide labor for the construction of infrastructure in the new city. In time, the ethnic and sub-ethnic groups, through their networks and associations, established themselves in occupational and trade specializations.[4]

While the British administration did not attempt to foster any form of 'nation-building instincts' in its subjects in Singapore, the legacy of the colony continues to affect Singaporeans till today. Not only did the British bequeath forms of democratic government to the state, such as the establishment of a Parliament, the judicial system and municipal structures; but also the education system, especially the use of the English language as the language for business and trade and a basis of 'neutral' communication among the races. The other enduring legacy was the classification of Singaporeans as Chinese, Malay, Indian, or Others, a practice that continued to foster a racial consciousness in Singaporean society.

The plurality of Singaporean society tended to be divergent in nature, and the laissez faire British administration did not try to foster harmonious relations between the groups. On the contrary, the British administration preferred to keep these as disparate, disunited groups. During the Japanese Occupation of Malaya and Singapore (1942–45), the Japanese capitalized on the lack of social capital in Singaporean society at that time, and used the Malay police against the Chinese and Indians (Quah 1990, p. 12). The Japanese were especially cruel to the Chinese population, as the Chinese had fought fiercely against the invading army. In one particularly notorious campaign, known as the *Sook Ching*, an estimated 5000–25,000 Chinese males between 18–50 years of age were massacred (Han et al. 1998, p. 22). The harsh treatment by the Japanese in the early days of the Occupation undermined their later efforts to enlist the support of Singapore people for the Japanese vision of a 'Greater East Asia Co-prosperity Sphere', which was to comprise Japan, China, Manchuria, and Southeast Asia (Library of Congress 1989).

When the Japanese surrendered in 1945, riots broke out in both Malaysia and Singapore before the returning British were able to control the tense situation (ibid.). This was a time when the certainty of British rule was replaced by widespread disdain for the colonial government, as the British could no longer resume their authority after having abandoned their subjects in wartime, as well as uncertainty about the future of the colony. This was a time when Singaporean society found itself ready for some kind of revolution in order to survive: in other words, the people of Singapore found themselves at the brink of the second founding moment.

---

[4]The smallest community in the late nineteenth century was the Indians, comprising Hindus, Muslims, Sikhs, and Christians. While the South Indians made a living as shopkeepers, laborers, stevedores, boatmen or bullock cart drivers, the north Indians worked as clerks, traders and merchants. The Malays were not very successful as traders and merchants, losing out to Chinese and European competition, and became shopkeepers, religious teachers, policemen, servants or laborers.

2.1.2 Rough Times and Independence

> …this one climacteric which triggered off a Singaporean entity. It started with the riots in 1964 when the police were out of our control and the army was not at our disposal; when we realized how vulnerable we were. So we learned to be patient but to be firm on gut issues – issues involving race, language, religion, culture. It is necessary to remind our young that when we started, in 1954 and when we formed government in 1959, we did not have the basic elements of a nation. The attributes of nationhood were missing: a common ethnic identity…a common language.[5]

The post-war transitional period marked the end of British authority and the move towards self-rule and independence for Singapore (1945–1965). The suppressed racial hostilities under the Japanese regime, and the dissatisfaction with British rule and the disruption of communitarian life were some of the reasons that led to riots and unrest in Singapore during this time.

Immediately after the Japanese surrender in 1945, anti-Japanese communists, who hid in the jungles during the war, embarked on a ruthless campaign to exterminate suspected Japanese collaborators (Han et al. 1998). The communists, organized under the Malayan Communist Party (MCP) took to the jungles again in 1948 after clashes with the British government, and fought the then government with guns and bombs in an armed insurrection called the "Emergency". Angry young people from the Chinese schools joined the communists to revolt against the British government.

There were also four riots linked directly to racial aggression, namely the Maria Hertogh riots of December 1950[6]; the July and September 1964 racial riots[7]; race riots in Singapore in 1965 blamed on Indonesian agitators, and the unpublicized racial riots which occurred in Singapore as a result of the spillover effects of the 13 May 1969 racial riots in Kuala Lumpur (Quah 1990, p. 58). During those and the following years, the Chinese in Singapore felt threatened because of the large Muslim presence in the surrounding countries (ibid., p. 79).

Chinese school students rioted against the government between 1954 and in 1956, not only on account of social justice but also for Chinese culture. In 1964, Malays rioted when the new housing program resulted in the destruction of their traditional *kampong* settlements (ibid., p. 7). These riots usually took place on the streets, involving violent clashes and sometimes damage to properties through fires. The Malay riots of 1950 and 1964 played out in large public open spaces such as the Padang and the open spaces at Empress Place.

---

[5]This is a quotation by Lee Kuan Yew from his January 1980 speech at the PAP 25th Anniversary Rally, titled "History is not made the way it is written". In the speech, he expounded his views on that which has shaped the Singapore entity.

[6]The Maria Hertogh riots occurred over a court decision to send a Dutch girl, adopted by Muslim parents and converted to Islam, to a convent, when her natural parents claimed rights to their child. Eighteen Europeans were killed and many more injured.

[7]These riots occurred when the print media reported the suppression of minority rights of the Malays, which led to suspicion and tension between the Chinese and Malays.

It was during this period of uncertainty that the political career of Lee Kuan Yew, eventually Singapore's first prime minister, took off. Lee was the founder and also the Secretary General of the People's Action Party (PAP) from 1955–1992. The PAP won the first general election in 1959 and every election since then, and in between 1968 to 1984, there were virtually no opposition members in Parliament.[8] Lee's motif was 'survival' right after Independence,[9] and this had actually never ceased to be part of his political rhetoric thereafter (Barr 2000, p. 32).

In order for the inchoate nation to survive, Lee devised, in 1965, the motif of the 'rugged' society, meaning strong effective government, uncorrupted civil service, a stable economy and political system poised to attract foreign investors (ibid.). The 'rugged society', however, was not like the western idea of rugged individualism or rebelliousness; on the contrary, this notion saw society as a whole being resilient, more as a 'herd' whose cultural instincts make them effective members of the collective. In other words, the idea was communitarian in nature (Barr 2000, p. 150). At the same time, the government clamped down hard on communal violence by dispersing Malay *kampongs* and curtailing Chinese activism.

2.1.3 Lee Kuan Yew, Values and Culture

In describing the processes through which the Singaporean public is shaped, it is useful to relate them to two defining themes of its developmental history—the leitmotifs of survival and multiracialism. Although the story of Singapore's development extends through complex trajectories, I believe that these are the dominant concerns influencing many others. One of the important sub-themes that emerge from the dialectics of survival and multiracialism is the notion of meritocracy—the idea that Singaporeans should be rewarded on the basis of competence, and not on aspects of their racial background or identity.

Singapore's recent history of founding moments implies that in the wake of each moment, a reminder of the need for survival re-surfaces. The nature of the survival leitmotif changes with time, along with the stage of development of the nation-state as well as the nature of the challenges facing Singaporeans. Meanings of survival have ranged from basic survival, as in living from hand to mouth; to survival as a people, as an independent country; to economic survival, as well as survival as a competitor in an increasingly globalized world. The various guises of survival in

[8]In the Singapore government, the Cabinet, led by the Prime Minister, controls legislative process. Matters of the state are debated in Parliament. The Cabinet is, however, the most powerful body that controls Singapore. Today, Singapore is still under one-party rule although a few opposition members have been elected into Parliament.

[9]Singapore gained self-rule from the British in 1959, but joined the Malayan Federation from 1963–65. The city-state became an independent republic in August 1965, when it became clear that it was no longer tenable to remain as part of the Malayan Federation.

the context of Singapore's history continue to shape the notion of the Singapore public.

What does 'multiracialism' mean in the Singaporean context? The term has been used in various ways to mean rather different things at different historical moments (Quah 1990). It sometimes refers to the equal treatment of the races by the government; at other times, it means that the various races should revert to ethnic heritage as inoculation against negative Western influences to Singapore's open society; it sometimes means that racial boundaries should be kept distinct, and that Singaporeans should organize themselves in racial communities for support (ibid., p. 16). Depending on the stage of cultural integration in the country, the meaning of a multiracial society had continued to change.

The concept of meritocracy is one borne out of the need for the survival of the multiracial society. As the notion of reward for good performance in order to ensure continued survival of the nation, meritocracy had been the unchallenged tenet of Singapore society, with the government and the people embracing the concept. To a great extent, the concept and practice of meritocracy had shaped the attitudes of the Singaporean public.

2.1.3.1 The Architect of Modern Singapore

Lee Kuan Yew is arguably the architect of modern Singapore. Politician and social engineer extraordinaire, he had been described by biographers variously as: a progressive, an elitist, a geneticist, a cultural evolutionist, a Confucianist, as well as a pessimist (Barr 2000; Tamney 1996). Although fundamental beliefs guiding his actions have remained steadfast throughout his political career, his ideas on shaping the nation and his ways of shaping Singaporean culture to achieve progress had changed—sometimes drastically, so that he seemed to be fraught with contradictions at times.[10] Above all, Lee believed that culture was malleable—a point which he freely admitted—which therefore could be used to shape people's habits, practices and ideology in order to achieve prosperity for the nation:

> Genes cannot be manipulated, right? Unless you start tinkering with it as they may be able to do one day. But the culture you can tinker with. It's slow to change, but it can be changed – by experience – otherwise human beings will not survive. If a certain habit does not help survival, well, you must quickly unlearn the habit…So I've got to try and get Singaporeans to emulate or to adopt certain habits and practices which will make Singapore succeed (Han et al. 1998, p. 179).

After stepping down from Prime Ministership in 1990, Lee became much more reflective of his life, such that in his 1998 authorized biography (Han et al. 1998), he sought to explain in his own words ideas and key experiences that shaped his view of the world. His changing notions, in particular, about the use of culture that

[10]Barr described him as: a socialist (from Cambridge days) who transformed Singapore into a capitalist economy.

followed him throughout his political career, are well documented in his numerous speeches and interviews with the media,[11] giving us a good picture of the man and his ideas.

Lee was highly aware that at the point of separation from Malaysia,[12] Singapore's resources were few and the survival of the nation depended on having a productive and dedicated workforce. As he prepared Singapore in the late 1950s to merge with the Malayan Federation, Lee leant towards an egalitarian "Malayan culture" that encompassed a blending of Chinese, Malay and Indian culture,[13] a rendition of multiracialism with Malay as the national language that would inoculate the large Chinese population from being influenced by a powerful communist China (Barr 2000, pp. 137–139).

However, after Independence from Malaysia, Lee adopted a western-style of governing and a western-oriented culture for Singapore society, with English as the medium of instruction in many schools and as the language of business. In fact, the Chinese in Singapore were often perplexed by the western style of governing of the PAP—it was only in the 1970s that "Asian values" again became incorporated as part of the PAP ideology (Tamney 1996, p. 8). The early 'melting pot' idea of a Malayan culture had been resigned to make way for a more pragmatic one, to allow Singaporean culture to develop over hundreds of years. Until then, and in order to insulate the citizens from becoming influenced by negative western cultural trends of the 60s, such as hippyism, casual attitudes to sex and drug-taking, the racial communities had to draw from their cultural heritage: a community-based "cultural ballast" to fuel the multiracial society towards progress (Barr 2000, p. 145). Lee's application of his ideas to shape Singaporean society became increasingly holistic to encompass more and more aspects of Singaporean life, often blurring the distinctions between public and private realms.

---

[11]There is a good selection in Han, Fernandez and Tan's *Lee Kuan Yew: The Man and His Ideas* (1998). See also Lee Kuan Yew, *Prime Minister's Speeches, Press Conferences, Interviews, Statements*, etc. *1959–1980*, Singapore: Prime Minister's Office. Also Lee Kuan Yew, *Senior Minister's Speeches, Press Conferences, Interviews, Statements*, etc. *1990*, Singapore: Prime Minister's Office.

[12]Singapore was part of the Malayan Federation from 1963–1965. The reasons for the separation will not be dealt with here, but one of the most fundamental reasons was cultural difference between the Malaysian polity and the Singapore politicians.

[13]"Although the ethnic categories were meaningful in the Singaporean context, each subsumed much more internal variation than was suggested by the term "race". Chinese included people from mainland China, Taiwan, and Hong Kong, as well as Chinese from all the countries of Southeast Asia, including some who spoke Malay or English as their first language. The Malays included not only those from peninsular Malaya, but also immigrants or their descendants from various parts of the Indonesian archipelago, such as Sumatra, the Riau Islands south of Singapore, Java, and Sulawesi. Indians comprised people stemming from anywhere in pre-1947 British India, the present states of India, Pakistan, and Bangladesh, and from Sri Lanka and Burma. Singapore's Indian "race" thus contained Tamils, Malayalis, Sikhs, Gujaratis, Punjabis, and others from the subcontinent who shared neither physical appearance, language, nor religion". (Excerpt from "Ethnic Categories" in Country Studies: Singapore 1989.)

2.1.3.2 A Culture of 'Campaigns'

The late 1970s and 80s were punctuated with numerous nation-wide campaigns to bring about desired habits, practices and values. These included the "Use Your Hands" campaign of the late 70s to inculcate the importance of blue-collar work to prepare the population for large-scale industrialization; the "Courtesy" campaigns, initiated in the early 80s with the modest aims of promoting courteous behavior towards tourists, have grown into more ambitious programs to promote a 'gracious society', spawning also in its wake, the "Kindness Movement".

Lee came to realize in the 1970s that the traditional cultures of Singaporeans (Chinese, especially) embodied an ethos that was useful for modernizing Singapore: they valued education and hard work, gave respect and status to state bureaucrats, and embodied faith that the state would be responsible for society's progress (Tamney 1996, p. 8). Lee set to devise his own version of Chinese culture, language and Confucianism as central features of Singaporean life (Barr 2000, p. 33). One of the most ambitious and enduring campaigns in Singapore's social engineering program must surely be the "Speak Mandarin Campaign", now called "Mandarin for Chinese Singaporeans". Initiated in 1979, the campaign was a project to replace the many dialects used by Chinese Singaporeans with the 'mother tongue'. The goals of the campaign were to improve communication between Chinese dialect groups and to promote Confucianism (Country Studies 1989). The surrender of the use of dialects was not only a language reformation, but also "implied a major transformation of the social structure of the Chinese community, because the associational and commercial structure of Singapore's Chinese-oriented society rested on (and reinforced) dialect distinctions" (ibid.).[14]

With this frame of reference, Lee's harnessing of Confucian values was seen as a move to ensure consensus and support for the extant government (Barr 2000, pp. 33–34). Lee put forward his argument that for Asians, the needs of the collective had always taken precedence over the individual:

> Nobody doubts that if you take me on, I will put on the knuckle-dusters and catch you in a cul-de-sac…Anybody who decides to take me on needs to put o the knuckle-dusters. If you think you can hurt me more than I can hurt you, try. There is no other way you can govern a Chinese society (Han et al. 1998, p. 126).

2.1.3.3 Neo-Confucianism and Shared Values

Lee's strong stance on how a Chinese society should be governed led many political commentators to view his promotion of Confucian values as a further

[14]Some academics also questioned the restriction of Chinese values to Confucianism and recalled that in the 1950s and early 1960s Chinese was the language of radicalism and revolt rather than of loyalty and conservatism.

justification to his authoritarian style (Han et al. 1998, p. 195).[15] Confucianism, as advocated by Lee at this point in Singaporean history, did not really appeal to Singaporean society. It was reworked into the more generic 'national ideology', elaborated as 'Asian values', and eventually approved by Parliament in 1991 as Singapore's "Shared Values".[16]

The four core values were:

- Community over self
- Upholding the family as the basic building block of society
- Resolving major issues through consensus instead of contention
- Stressing racial and religious tolerance and harmony.

The ideal society, as spelled out in the Shared Values, was one in which individual happiness and needs were subsumed under the collective good. As the family was recognized as the basic unit of society and not the individual, the overall happiness of the family outweighed benefits to individual members of the family. Divorce, single parenthood and singlehood were frowned upon—the family should not be sub-atomized.[17] Regarding the favoring of consensus over contention, critics had argued that the government simply would not tolerate criticism and contention, and that consensus meant going along with the government on all issues. But as Lee had said:

> ...My job is to persuade my flock, my people, that that's the right way. And sometimes it may be necessary not to tell them all the facts because you will scare them. What the crowd thinks of me from time to time, I consider totally irrelevant...The whole ground can be against, but if I know this is right, I set out to do it, and I am quite sure, given time, as events unfold, I will win over the ground... (Han et al. 1998, p. 229)

2.1.3.4 Media and Censorship

There had long been a criticism of the Singapore mass media, especially from foreign media, that it did not put forward independent views on national issues (Tamney 1996, p. 61). With the legal structure supporting restrictions placed on publications that did not work within the realm of non-contention,[18] outspoken

---

[15]Singapore society is considered by Lee as a Chinese society, and he makes little pretense to hide this view.

[16]A National Ideology Committee was established and headed by the younger Lee, which produced the "Shared Values" report in 1991. Drawing from Chinese, Malay and Indian cultural traditions, the national ideology was drafted to counter undesirable traits from the West, namely individualism, self-centeredness, not working hard, and being suspicious of political leaders (and thereby limiting their powers), casual sexual relationships, and single parenthood.

[17]These values are also embedded in the rules for purchasing public housing flats.

[18]Under the Newspapers and Printing Presses Act (NPPA), passed in 1974 and amended in 1986, the government could restrict—without actually banning—the circulation of any publication sold in the country, including foreign periodicals, that it deemed guilty of distorted reporting.

foreign publications such as the *Asian Wall Street Journal* and *Time* magazine's Asian edition had suffered suspension of circulation within Singapore in 1987. The government had also restricted the circulation of *Far Eastern Economic Review* and *Asiaweek* in 1987 for "engaging in the domestic politics of Singapore".[19] At the 40th World Congress of Newspapers Publishers held in Helsinki in 1987, then Minister of Trade and Industry Lee Hsien Loong justified the action against foreign press:

> The reason Singapore is so concerned about foreign press involvement in domestic politics is that we have seen how the media may bring in undesirable values, how newspapers can be used to carry out covert subversion, and how inflammatory reporting can lead to racial riots (Quah 1990, p. 59).

However, it did surprise some of the elder Lee's biographers that Singaporeans rated press freedom lowly on their list of priorities and did not find it particularly imperative that the press should provide a check on government—foremost on most Singaporeans' minds was that the government should bring wealth and progress to the nation (Han et al. 1998, p. 217). In other words, Singapore, in the early years, exemplified the "developmental state"—a condition in which "the state manages the economy and is relatively free from having to meet short-term populist demands" (Tamney 1996, p. 59).[20] Lee had anticipated the political sentiment of the time and advocated an 'Asian' style of democracy when he said in 1991:

> Simply modeling a system on the American, British or West European constitution is not how Asian countries will or can go about it. The peoples of Asia want higher standards of living in an orderly society. They want to have as much individual choice in lifestyle, political liberties and freedoms as is compatible with the interests of the community. After a certain stage of advance in education and industrialization, a people may need representative government, however chosen, in order to reconcile conflicting group interests in society and maintain social order and stability. Representative government is also one way for a people to forge a rapid consensus, a social compact, on how a society settles the trade-off between further rapid economic growth and individual freedom.[21]

2.1.3.5 Asian and Modern

Singapore's national ideology was conceived to maintain Asian societal norms. But this ideology, especially from the view of those from the 'outside', is also "radical

[19]This information can be referenced to the Ministry of Communications and Information 'Press Release' page on their website 'http://www.mci.gov.sg/web/corp/press-room/categories/press-releases/content/circulation-of-foreign-newspapers-in-singapore'.

[20]Singapore is often described as a developmental city-state, in that "it establishes as its principle of legitimacy its ability to promote and sustain development, understanding by development the combination of steady high rates of economic growth and structural change in the productive system, both domestically and in its relationship to the international economy". In other words, economic growth is prioritized above all else as an indicator of government performance.

[21]This quotation is from Lee Kuan Yew's address at the Ashahi Shimbun symposium, on May 9 1991. Also cited in Han et al. (1998), p. 372.

because it is such a basic attempt at social design: to create a culture that is not Western, is modern, and is Asian" (Tamney 1996, p. 19). Singapore's leaders have never described themselves or the form of government as authoritarian (a term commonly bandied by foreign media) but have continued to defend their style of government as a necessarily Asian style of democracy, steeped in Confucian norms (Tamney 1996, p. 57). In fact, Lee had described himself as a liberal:

> Today, I would describe myself as...between socialists and conservatives. I would put myself as a liberal. As someone who believes in equal opportunities so that everybody gets and equal chance to do his best and with certain compassion to ensure that the failures do not fall through the floor...A liberal in the economic sense of the word...Not a liberal in the sense of the American word "liberal"...But a liberal in the classical sense of the word, in that I'm not fixated to a particular theory of the world, or of society. I'm pragmatic. I'm prepared to look at a problem and say, all right, what is the best way to solve it that will produce the maximum happiness and well-being for the maximum number of people, You call it whatever you like (Han et al. 1998, p. 130).

The last of the four points of the Shared Values stressed the premium placed on racial and religious harmony. This remained quite a powerful motif in light of past racial and religious conflicts—discussions in the public realm bordering on race or religion issues treaded, so to speak, on sacred ground and were quickly extinguished when they were deemed to be harmful to the harmonious co-existence of the different groups.

2.1.3.6 Ethnic Identity

The ethnic categories, referred to as 'race' in the Singapore context, continued to be a conscious organizing structure of society, such that beyond the family, the 'race' community formed the next level of support for the individual. The Singapore identity was defined as membership of the 'natural' groups of Chinese, Malay, Indian or Others, assuming that these were homogenous groups and that the boundaries between the races remained indefinitely (Country Studies 1989).

The government, right from Independence, had worked to break up ethnic enclaves, as these were believed to be the breeding grounds for racial incitement and communitarian violence, and to reshape communities through the distribution in public housing and through community centers throughout the island-state. However, the consciousness of racial difference continued to be reinforced in the government's specified racial quota of residents for every block of public housing flats, and celebrated through different public holidays allocated to the different races and religious groups within the calendar year. Nevertheless, although Singaporean society had a "high level of racial consciousness", it had a "low level of racial tension" (Barr 2000, p. 205).

Ethnic identity played out in everyday life through community structures such as the mosque, sharia (Muslim law) courts, Hindu temples, as well as religious bodies. Historically, the Chinese in Singapore had the densest network of ethnic associations, achieved through clan organizations, and business,[22] religious and recreational associations.[23] Membership of Chinese clans used to be restricted to those speaking the same dialect, or those who could trace their ancestry to the same region of China. However, clan memberships have since declined, as younger generations of Singaporeans have little affiliation to dialect group (thanks to years of "Speak Mandarin" campaigns).[24] They were not aware of their ancestral origins in China, or were indifferent to these. Since then, these associations have changed in form and content, promoting Chinese culture in general. They have also been active in the larger community by organizing charitable events or taking up endowment and advisory and sponsorship roles to the schools, scholarship funds, etc.—as did the Ngee Ann Kongsi, one of the most influential of such associations in Singapore.

2.1.3.7 Social Engineer Extraordinaire

By the end of the 1980s, most Singaporeans would describe themselves as middle class,[25] owning a HDB flat with secure savings in the Central Provident Fund; having a steady income (or two incomes, with a working wife) and a well-stocked home; and spending time on hobbies, sport or holidays abroad. They would probably use public transportation, though many would still like to own a car (Country Studies 1989).

Lee had no doubt that prosperity had to come through hard work and social discipline, taking always the best available course for economic growth without letting ideology, religion or cultural prejudice exclude any option for achieving economic success (Tamney 1996, p. 10). He would also not hesitate to create the social conditions to favor such growth, making him social engineer extraordinaire. It was a national obsession to be ahead in rankings for business performance, service, education, efficiency, and other kinds of indexes, especially if these had impact on economic growth. Despite the perceived authoritarian approach to

[22]The Singapore Chinese Chamber of Commerce and Industry, founded in 1906, was the overarching association that represented the entire Chinese business community. A federation, its constituent units were not individuals or individual businesses, but associations. Its basic structure consisted of representatives of seven regional associations (Fujian, Teochiu, Cantonese, Hakka, Hainan, and "Three Rivers") and ninety-three trade associations, each one usually restricted to speakers of one dialect.

[23]According to Country Studies: Singapore, these number more than 1000 still, in the late 1970s.

[24]Up till the 1980s in Singapore, many Chinese-run businesses still fell into the traditional occupational groups particular to the different dialect groups. For the proprietors and employees of many small and medium Chinese businesses, it continued to be fruitful to identify with the powerful clans and associations of the dialect and sub-ethnic groups, as the social solidarity proved economically advantageous. This effect has to a certain extent been eroded by the rise of multi-national corporations in Singapore in the 90s, and the powerful government-linked corporations that encompass many industries and businesses in Singapore.

[25]In 1987, then Prime Minister Lee declared Singapore a 'middle-class society', based on the criterion that more than 80 % of Singaporeans owned the property they lived in.

politics, the superlative transformation of Singapore's economy, society and physical environment, and the corruption-free government, attest to Lee's leadership and visionary skills.

2.1.4 'The Next Lap': The 1990s, 2000s and Beyond

When asked to name the most important invention of the 20th century, Lee Kuan Yew had, without reservation, singled out the air-conditioner. The air-conditioner, a machine enabling Singaporeans to remain cool in Singapore's humid tropical weather, is emblematic of the Singapore government's desire to provide economic growth and material comfort for Singaporeans. This desire is reflective in the adaptive stance taken by the government in its adoption of economic strategies that changed along with the economic climate of the times.

In the 1980s, when Singapore's growth was on the fast track, the promotion of the Shared Values as a way of life was widely accepted in exchange for the steady growth of the economy, and the stable environment of the city-state. However, especially in the wake of the post-1997 Asian economic downturn, these 'values' were transformed. In place of these Shared Values, the new horizon of Singapore Vision 21 (Century) became another new 'founding moment', so as to prepare the ground and the workforce to populate a "knowledge-driven economy". As Singapore strode into the 1990s and the increasingly globalized realms of economic, political, and social relations between borders and domains, there was a "state transition", a form of adaptation of the government and its political structures from "authoritarian taskmaster" to "venture capitalist" (Ong 2004, p. 178).

> Entrepreneurship and innovation will be key ingredients of economic success. The more developed we become, the less we can merely follow the path blazed by others. Singaporeans need to venture forth on their own, to grow activities in the region, to create and develop key product and knowledge niches to maintain our competitive edge.[26]

With 50 years of economic development since its independence, Singapore now had a maturing economy but its achievements had been in many ways state-led. The government had played a major entrepreneurial role through the setting up of many government-linked companies (GLCs) such as the national airline, telecommunications provider and the creation of Jurong Island as the new chemical hub of the city-state. Innovation and enterprise, mostly from private sector initiatives, were needed to build the engine of further growth in an increasingly competitive and globalized world economy.

This required some kind of strategy to both attract and retain international talent and expertise, as well as for the population to remain "socially cohesive and nationally resilient" (Ong 2004, p. 178). A new kind of Singaporean was needed—one who was adventurous, a risk-taker, independent and innovative—an

[26]This quotation is by then Deputy Prime Minister, Lee Hsien Loong, in 1998.

entrepreneur. Public discussion had been nudged from the need for discipline and good work ethics to the need for creativity and risk-taking (Tamney 1996, p. 70). Schools were in fact criticized for producing risk-adverse conformists at a time when the economy needed innovators and entrepreneurs (ibid.). Singapore needed these on top of the hardworking, frugal, disciplined and loyal worker. However, political commentators have observed that it would be naïve to imagine that these values should be confined to the workplace, and that traditional "Asian values" would still prevail within family and society (Ong 2004, pp. 185–186).

> The Singapore that marches into the 90s is not the same Singapore that was thrust into independence in the 60s. The electorate is more affluent, more secure, more confident, more educated, and less tolerant of governmental paternalism, however well meant that paternalism may be (Woon 1991; Tamney 1996, p. 69).

Clearly, these changes to the make-up of the Singaporean would require a different style of governing, allowing independent-minded and well-educated citizens to take control of their own lives and even to be able to criticize public policies. It was not difficult to anticipate that, "(t)he desire to be independent, once fixed in personalities, cannot be limited to only economic matters… The new Singaporean is less content with a paternalistic, pseudo-Confucian style of governing. This state of affairs is not a result of capitalism itself but of the stage of capitalism that Singapore is entering" (Tamney 1996, pp. 70–71).

2.1.4.1 The New Guard

Lee handed over the premiership of Singapore to his successor, Goh Chok Tong, in 1990—a changeover from the Old Guard, who had nurtured the state since its independence, to the New Guard, who would bring it into the 'Next Lap'.[27] Since 1985, moves had been made by the government to be more inclusive (Tamney 1996, pp. 71–74). These included the setting up of the Feedback Unit in 1985, which collects opinions, views and suggestions from the public in writing and through public forums; the introduction of Nominated Members of Parliament (NMPs) in 1990, who have limited voting privileges and act as 'sanctioned opposition', representing varied interests and groups such as youth, women, and the environment; expanding the role of the elected President in 1991 with powers over the use of state fiscal reserves and veto powers in the appointment of top civil servants; and the establishment of Town Councils in 1989 by way of decentralizing the control of new towns—these constitute mainly maintenance activities.[28]

While not actually letting up on economic development, Goh Chok Tong's government of the 90s was committed to the concept of a 'caring nation'. One of the ways of achieving the 'gracious' society and 'caring nation' was through the development of the arts in creating a cultured people. Culture was also deemed to

[27]The term Goh coined to mark his term of office.

[28]Council members are, however, not elected by members of the new towns.

be an area of viable economic growth. It was also acknowledged that Singapore needed to improve upon its image of being culturally sterile, so as to persuade expatriates to work and stay in Singapore, and at the same time, prevent a brain-drain of culturally-minded Singaporeans. The defining difference of Prime Minister Goh over Lee was his emphasis on the building of a 'gracious' society:

> If Singaporeans are rich but crude, rich but selfish, rich but uncaring, the society cannot hold together for many years because we are going to have internal conflict, tension, and very quickly, the whole place will fall apart (The Straits Times 1996).

Under the New Guard, the importance of social order still preceded the expression of individual views: street protests were still outlawed, and assembly of over five persons in public spaces still required a license or permit to be obtained. In other words, there was a move towards popular participation in the political process, but in a controlled manner.

2.1.4.2 Beginnings of a Civil Society

The government continued to co-opt brilliant and talented Singaporean elites into its folds, including those who had no political ambitions but were willing to head statutory boards, organizations and trade unions. Through the leadership of these non-political bodies, the influence of the government percolated deep into the various structures of society, forming a collaborated civil society. The PAP, the dominant political party since independence, did not have credible opposition leaders. Opposition parties in Singapore failed to attract good candidates, representation of minority groups, or resources for successful election campaigns (Tamney 1996, p. 63).

Another reason for weak civil society was that the dominant Singaporean Chinese society was built upon low social capital due to the numerous dialect groups, as well as the fact that Chinese enterprises were traditionally family-run businesses, with the *kongsi* being the association serving the dialect group. The realms of trust extended first within the family, then the dialect groups or clan association, and then to the larger Chinese community. The existence of Malay and Indian ethnic groups further compounded the problem of a 'low trust' society. The government's initiatives, including the use of common languages and reinforcing the tenets of the multi-racial society as well as religious harmony were based on building a harmonious nation while preserving ethnic identities within a heterogeneous society.

2.1.4.3 A Multicultural Society

Although standardized education and equal treatment of the different races by the government have resulted in more shared attitudes between the different races, this shared existence is one of proximity than of a real 'melting pot', as underscored in a reflective speech in 2004 by the then much-mellowed Minister Mentor Lee Kuan

Yew, made in response to a Parliamentarian's argument for the emergence of a Singaporean consciousness that would transcend ethnic and cultural urges. He believed that ethnic identities are deep-rooted and it would take a long time before people felt Singaporean above their own racial identity:

> …he made his point with a stark scenario that, in multicultural Singapore, is increasingly commonplace. "Ask yourself this question: if you have a child and he brings back a boyfriend or a girlfriend of a different race, will you be delighted?" "I'll answer you frankly - I don't think I will. I may eventually accept it," he said (The Straits Times 2004a, b, c, d, e, f, g).

2.1.5 Political Trajectories

With the handing over of the premiership of Singapore in 2004 to Lee Hsien Loong, the current Prime Minister, change seemed imminent, but not in the form of a revolution of the political culture that had prevailed since the inception of the city-state. In an interview with *Fortune* magazine, the younger Lee had promised to "cut the apron strings of a nanny state"—he cited the already lifted longstanding bans on bar-top dancing and bungee jumping as signs that the new regime was "prepared to take the plunge and make even deeper changes in society" (Kraar 2004). Lee even urged Singaporeans to speak freely, as "(d)isagreement does not imply rebellion" (ibid.). As Lee pointed out in his interview, this vision of freedom would more likely entail citizens' participation in the shaping of public policy, with the ultimate decision still made by the government.

In 2004, the elder Lee assumed the role of Minister Mentor—a position specially created for him—to serve as an advisor during a period of political transition. Lee remained an influential figure in the background from frontline decision-making until his retirement from politics after the 2011 General Election. Lee Kuan Yew's passing in 2015 immortalized him as a national hero and he was heralded as a visionary among the founding fathers behind Singapore's success story. As his son and Prime Minister Lee Hsien Loong declared in his eulogy, "To those who seek Mr. Lee Kuan Yew's monument, Singaporeans can reply proudly: 'look around you'".[29]

2.1.6 Cultural and Social Trajectories

In the mode of continually adapting the course of action to deal with an ever-changing world, a new wave of structural changes embodied in "Remaking

[29]This quotation is from Prime Minister Lee Hsien Loong's eulogy for his father, Lee Kuan Yew. Source: http://www.pmo.gov.sg/mediacentre/transcript-eulogy-prime-minister-lee-hsien-loong-funeral-late-mr-lee-kuan-yew-29-march.

Singapore"—a series of structural changes that would support a more competitive city-state in the age of globalization—was proposed in 2003:

> We have concrete ideas on how to deepen the ties of Singaporeans to Singapore, and to each other. We have signaled that globalized Singaporeans must be considered full-fledged members of our society. We have proposed important changes to our education system to help meet the aspirations of Singaporeans and make us a more adaptive people. There are specific proposals aimed at nurturing a culture of expression and participation, of graciousness and compassion. We also point out particular policy shifts that would equalize the treatment of Singaporean women and men, and recommend how to strengthen family life, deepen our heritage roots and make Singapore a fun place to live in.[30]

In addition, the report of 2000 for turning Singapore into a "Renaissance City" aimed to transform Singapore into a "highly innovative and multi-talented global city for the arts and culture".[31] This ambition served to animate existing public spaces as sites for artistic expression. In addition, the proliferation of arts venues such as the Esplanade—Theaters on the Bay, as well as the establishment of Marina Bay, brought with them ancillary public spaces that increased the variety and choice of public spaces in the city, and enhanced the overall vibrancy of the city.

The trends already in place will continue, such as the increasing number of foreigners in Singapore, at either end of the global flow of labor. There is also an observable trend in the increasing vocalization of 'multiple publics' such as women's groups, LGBTs, foreign workers, nature protectors, and consumer groups. One way to see this is that there is an emerging critical public, which, empowered as organized groups, are able to articulate grievances more persuasively than if they had acted as individuals in voicing their complaints. These groups have continued to push for representation and for their causes in different mediums of public representation, one of which is, of course, public space.

Organized activism, such as the 2004 organized 'protest' of breast-feeding mothers at the Esplanade, the 'Pink Dot' event held every year since 2009 in support of the Lesbian, Gay, Bisexual and Transgender community in Singapore, and organized 'spontaneous' behavior in public such as flash mobs, a phenomenon seen in other major cities of the world, have made their appearances in public spaces in Singapore. With the increasing tolerance of the state towards the

---

[30]This quotation is from a speech by Dr. Vivian Balakrishnan, Chairman of the Remaking Singapore Committee, at the Remaking Singapore Report Presentation and Appreciation Lunch, 12 July 2003. Its report, submitted to Prime Minister Goh in July 2003, proposed recommendations around four themes for renewal and change, viz. A Home for All Singaporeans, A Home Owned, A Home for All Seasons, and A Home to Cherish. 2. Collectively, these recommendations aimed to (i) unleash the creative potential of Singaporeans and attract discerning talent to live and work here; (ii) fine-tune Singapore's safety nets and engender a more compassionate society; (iii) expand common spaces and strengthen social cohesion; and (iii) create more opportunities for Singaporeans to contribute to the country and strengthen their emotional bonds to Singapore (Apr 2004).

[31]The report outlined two main aims: to establish Singapore as a global arts city and to provide cultural ballast in the ongoing nation building efforts, and was endorsed by the government, to be implemented over five years.

expressions of views in public, a move towards less 'muted' forms of spatial practice in public space is perhaps in the making. There are also signs of an emerging and active civil society in the form of independent interest groups who discuss issues in public and create awareness of the issues at stake, and then put forward their argument to the state through persuasion instead of outright dissent.[32] It is also a fact that with the availability of the Internet and the free flow of information that many Singaporeans now have access to, information can no longer be controlled by the so-called 'nanny state'.

At a speech to University students in 2003, the younger Lee emphasized four major challenges facing the nation that have persisted into the 2010s.[33] The first focused on maintaining racial and religious harmony, as it was observed that Singaporeans have become more religious and tended to form friendships and associations within religious groups, leading to less interaction with people of other beliefs. Friendships also tended to be formed within limited racial communities.

The second dealt with international terrorism and the threat posed by extremist groups in the region. With the arrest within Singapore of members of extremist cell, Jemaah Islamiyah,[34] it was inevitable that a general climate of insecurity prevails, and the local Muslim population is regarded with some degree of suspicion. Singaporeans are also constantly reminded of historical events involving antagonism with its Muslim neighbors: the 1964 racial riots, the Confrontation with Indonesia, and the separation from Malaysia. Lee especially warned against the forming of racial enclaves within the 'private' domains of the home communities, at work and in schools. Talk regarding these issues is encouraged within government-initiated focus groups called "Harmony Circles" and "Inter-Religious Confidence Circles". The issue of international terrorism becomes more acutely important in 2016 with the rise of the Islamic State and religious fundamentalism around the world.

The third challenge involved the building of social cohesion, as he observed that the opportunities offered by globalization might not benefit Singaporeans equally, and that Singapore society would become more stratified. He advocated the upholding of meritocracy as the paradigm to prevent a divided society. Prime Minister Lee Hsien Loong echoed this sentiment in 2015, when he warned that Singapore must be wary of "dissolving into globalization" and emphasized "strengthening the national identity" (Today 2015).

The fourth challenge was political, in that he felt Singaporeans needed to be "informed, engaged and committed to Singapore", and that there was a need for "broad, political consensus" in building strategies for Singapore's survival. He

---

[32]One such example is 'The Green Corridor' preservation project, which was successful due to the persuasive efforts of The Nature Society, which put together a proposal for the conservation of a 173.7 ha rail corridor running through the island.

[33]The speech by the Deputy Prime Minister Lee Hsien Loong was at a Pre-University Students' Seminar in June 2003, called Singapore@WAP (Singapore at Work and Play).

[34]Members of the group were accused of hatching a plot to sabotage American military interests in Singapore.

pointed out though that this should be achieved via "national education", i.e. through information and persuasion, instead of "indoctrination".

While Lee spoke in 2003, it is interesting to observe the parallels in his thinking, in relation to the elder Lee's, who had socialized Singaporeans to the idea of a multiracial society, Shared Values, meritocracy, and the value of "consensus" for Singapore's continued survival, some thirty years ago. Albeit, the style of the younger Lee's delivery was more persuasive than the allegorical "knuckle-dusters" approach of a generation ago.[35]

2.1.6.1 Assimilating the 'New Normal'

The 2011 General Elections was heralded as a 'new normal' for Singaporean politics. The elections ushered in a more contested political space and a more demanding electorate, as evidenced by debates over developmental trade-offs such as the Bukit Brown Cemetery, the demolition of Rochor Centre, as well as the environmental impact of the Cross-Island Line MRT running under the Central Nature Reserve. Even though the 2015 elections were cautiously described as a return to more established norms (The Straits Times 2015a, b, c), it was clear that wider social media participation and new forms of citizen engagement had established their roots in the broader perception of the Singapore public—especially with a new generation of voters growing up post-Independence. The 2013 Population White Paper reported Singapore's birth rate as steadily falling over the last three decades, with its total fertility rate sinking under the optimum replacement rate of 2.1. Singapore's life expectancy was also reported to have increased from 66 years in 1970 to 82 years in 2010. Falling birth rates coupled with increasing life expectancies have ultimately resulted in an aging and shrinking citizen population and workforce. To mitigate this, Singapore has put in place enhancements to policies that encourage marriage and parenthood, and a friendly immigration policy (for both for high and low-skilled labour) to add diversity and age-balance to the citizen population (NPTD 2013). Singapore accepts approximately 15,000 and 25,000 new citizens per year to supplement its citizen population, and by 2030, Singapore would have to plan for a population scenario of 6.9 million (ibid.). Despite the balanced recommendations of the paper, emotive debates over the erosion of the Singapore core resulting from the inevitable need to increase the foreign workforce spelt the need for government agencies to engage the citizenry much more than in the past in enacting development policies for the future.

These challenges would appear to benefit from an increased publicity of Singaporean life, so as to bring into the public domain racial relations, and bring to public discourse the previously tabooed issues of racial harmony. The concept of meritocracy makes visible the achievements of individuals, while the emphasis on

[35]The elder Lee made a speech in which he said that he would not hesitate to confront dissenters with knuckle-dusters in a back-alley.

consensual politics through information and persuasion would necessitate bringing increased levels of political discussions and debate into the public realm. Public space may indeed be one of the avenues to provide the publicity necessary to articulate these challenges and become the site of spatial discourse towards a more pluralistic and adversarial 'consensus', rather than a homogenization or glossing over of differences to create an appearance of harmony. The emergent public space-in-the-making could be a contributing force towards a new standard of transparency in democratic process.

2.2 A Brief History of Urban Public Space in Singapore

The nature of government in Singapore, the pragmatic economic emphasis on many fronts, the adoption of neo-Confucian ethics in the forging of national values and the phenomenon of a national middle-class constitute a particular political and social milieu in which the production of public spaces has taken place in the last few decades. However, the provision and use of public space in Singapore shifts through a history tied inadvertently to national ideology, national development and control exercised by the state.

Far from the sedate image conjured up by the notion of Singaporean public space, the history of such had been as traumatic and crisis-loaded as any in many developing countries in uncertain and trying times. The particular type of agoraphobia exhibited by the state in relation to the provision of public spaces had its roots in such a history, which has had a lasting impact on the built environment of Singapore. The range and types of public spaces that emerge within the rapid urbanization of Singapore is tied as much to its social history as to its national aspirations, etched onto the urban template.

2.2.1 Public Space in Singapore: A Short History

2.2.1.1 Lawns, Far from the Madding Bazaar

The compartmentalization and ordering of districts in the nineteenth century plans for Singapore provided the colonial administration with the desired segregation of the European Town and the native quarters. Public spaces included lawns like the Padang, promenades like the Esplanade or Queen Elizabeth Walk (as it was later named), and squares like Commercial Square and Empress Place in the city center. The Padang[36] served as the visual foreground for the important municipal buildings

[36]Originally named the Padang Besar, (*padang* is the Malay word for "playing field") it is the open green space of the British colonial plan for Singapore, as the *maidan* (Bengali for "field") is to Calcutta (now Kolkata).

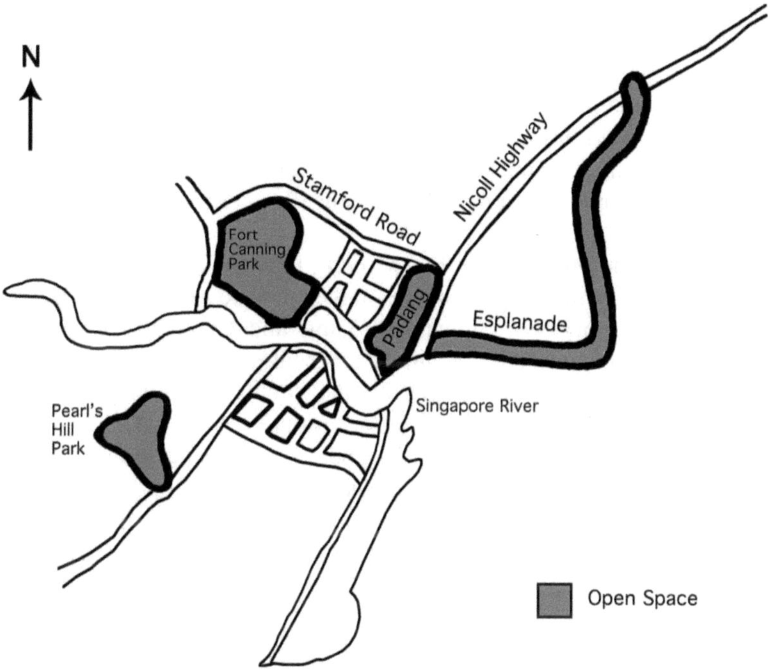

Fig. 2.1 Open space in the Raffles Plan. *Source* Author's Collection

of the British administration. It was also spatially the social center for colonial life in Singapore, forming the nexus of European public space from the Esplanade to Fort Canning Hill. Ironically, the Padang was the site of assembly of the POWs after the surrender of the British to the Japanese during World War II in 1942. The Australian and British troops were subsequently made to march 22 km to their internment in Changi Prison (Fig. 2.1).

The Padang also served as the symbolic venue of the surrender of the Japanese a few years later in 1945[37]—with an elaborate victory ceremony and celebratory march-past, culminating in the public punishment of the Japanese prisoners-of-war who were made to dig trenches on the Padang, which served no real municipal purpose. Following soon after, the Padang was once again the venue of the victory rally of the first elected legislative council of the new self-governing nation in 1959.

The open, green spaces of colonial Singapore served not only as foreground, but also as separators of the ruling class from the native and immigrant population of the island. While large compounds and high fences surrounded residential properties of the colonials, the urban substance of the ethnic wards in the city consisted of cramped city blocks of two to three storey 'shophouses'—the arcaded, terraced blocks with commercial usage on the ground level, and residential quarters 'above

[37]The Japanese occupied Singapore from 1942–1945.

the shop'. Public spaces in the crowded bazaar-like conditions of the wards consisted of the streets, verandahs or 'five-foot ways' and the back alleys of the shophouses. Activities of everyday life spilled onto the streets as much as they were confined to the interiors, such as eating, sleeping, gambling, selling, buying, playing, praying, weddings and funerals. The blurred boundaries of public and private spaces in the shophouse districts were often the site of contestations of rights between enforcers of colonial rules and the defiant immigrant population (Yeoh 1996) (Fig. 2.2).

Entertainment and amusements for the early nineteenth century Chinese immigrants were primarily religious in nature, rather than to do with commercial consumption evident in modern Singapore (Yung and Chan 2003, pp. 153–154). Examples of these include the staging of different genres of Chinese opera, performed on temporarily erected stages on the streets in Chinatown and by the Singapore River, on festivals and special occasions.

The street theater, held in honor of the gods, were often also the site of deviant activities such as gambling, murders, fights and elopement—and so were occasions when women were discouraged to be in public, to stay away from the predominantly bachelor immigrant workers (ibid., p. 158). Puppet theaters and public storytelling were also important events of the social calendar of the time. Until the

Fig. 2.2 Chinese Wayang, circa 1950s. *Source* Ministry of Information and the Arts Collection, courtesy of National Archives of Singapore

Fig. 2.3 Maria Hertogh riots, 1950s. *Source* Kenneth Chia Collection, courtesy of National Archives of Singapore

1940s when leisure activities moved within amusement parks, theaters, dance halls and cinemas, public spaces such as streets and alleys served the multi-faceted function of social spaces, activity focal points and cultural repositories.

2.2.1.2 Taking to the Streets

Post-war Singapore was a time of transition that was marked by flashes of violence, rioting and street protests. The uncertainty of the future after the Japanese surrender, the resentment towards the colonial government, the Communist insurgency and the rising of previously suppressed racial tensions to the surface, were all contributing factors to the outward expressions of anger. There were street protests and riots, which stemmed from racial and religious conflicts, anti-British and workers' protests, as well as those arising from perceived cultural discrimination,[38] all taking place in public spaces such as in the streets and at the Padang. Visually

[38]These included the Maria Hertogh Riots (1950), the anti-British riots and street curfews, the Hock Lee Bus workers riots (1955), the race riots of 1964 (which led to the expulsion of Singapore from the Malayan Federation) and the Chinese students' rallies of 1965.

well-documented as a period of insurgency, the period of turbulence had been used by the state as the dominant discourse to portray the defiance of authority as instances of chaos and loss of control (PuruShotam 2003, p. 47) (Fig. 2.3).

An alternative reading of the series of events may yield instances of civil society in action—attempts by the "body politic" to shape governance, or to "un-silence" repressed views (ibid.). These expressions of defiance challenging the dominant view in public space had been cast as representative of an undesirable period of Singapore's history, not to be repeated by citizens under a democratically elected government. These images have been evoked as the reason for not condoning public protests. These instances of insurgency and public shows of anger made a deep impression in the minds of Singaporeans, and had important effects on the future of the provision of public spaces in Singapore. It is an unwritten but well-known fact that subsequent planners of public space in Singapore had in mind not to provide large spaces for public gathering—as is the case in planning for urban renewal in the city, in the design of new towns and in the planning of the University of Singapore (which was completed in 1977).

The 1950s and 60s were an interesting era from the point of view of the Padang, which became a 'people's place' and a 'site of representation'. It became the venue of symbolic confrontation between the masses and the government, in the case of the *Nantah*[39] students' rally and the march of banned societal groups. At the same time, the space was 'appropriated' socially, with roadside stalls accessible to the masses such that the Padang functioned not just as a symbolic venue, but also as a social venue in the city.

2.2.1.3 From the Streets to the 'Garden City-State'

> Socialism had been born in crowded conditions – the street, the tenement, and the dormitory – and, in order to continually recreate itself, it needed to architecturally reproduce community…the street and the square, not the house, was the natural environment for the politics of progressive reform…By the time the Left "took to the streets" of Singapore in a last bid for power in the mid- and late- 1960s, "the street" was in the process of disappearing (Clancey 2004, p. 44).

The urban renewal policies of the 1960s and 70s basically led to the enforced mass exodus of the population from the central areas to the peripheral new towns, so that the Central Business District could be developed as the commercial heart of the city. There were also attempts to break up potentially insurgent communities who lived in squatter camps and *kampongs* in the central areas. The residents of such communities were often thought to be recruited by Communist insurgents, or were ethnic enclaves prone to acting within ethnic interests or inciting religious unrest (ibid. p. 45). So while Singapore's public housing programs had the official rhetoric of "building communities", the early housing of the 1960s tended also to split up

[39]Nantah is the shortened name of Nanyang University, the only Chinese university in Singapore, that was widely used in local lingo.

Fig. 2.4 Toa Payoh New Town, circa 1960s. *Source* Ministry of Information and the Arts Collection, courtesy of National Archives of Singapore

"pre-existing communities of the same race" to prevent the formation of ethnic enclaves (ibid., p. 45). However, policy-wise, the concept of distributing the races even at the level of each block of flats was crucial in the building of a multiracial society and achieving social integration.

Spatially, the street-based city blocks of the central district and the close association of many squatter camps with the street were being replaced in the new town environments by open, green areas between buildings. Blocks of flats no longer front the street, such that the once close relationship of living spaces and the street became dissociated. Elsewhere in the city, the informal economy of street vendors, hawkers and other temporary inhabitants of the street were evacuated to sanctioned accommodations, such as in markets and food centers, with proper licensing and sanitation facilities. Streets were widened to ensure smooth traffic flow and served as transport arteries. They were aesthetically re-modeled with landscaping and footpaths in line with the imaging of the 'Garden City'.[40] Public space in the city was the visual embodiment of the modernization of Singapore, and metaphorically announced the arrival of "clean government, clean civil service"[41] (Fig. 2.4).

[40]The Garden City concept here is not the notion as proposed by Ebenezer Howard. The term 'Garden City' in Singapore was more apt to mean the literal introduction of plants into the city to procure a green mantle and a pleasant environment.

[41]This was a catch phrase of the 1970s.

The separation of home and the city saw some significant changes. Public space provision in the central area was based on a framework devised in 1972[42] by the United Nations Development Program (UNDP) and adapted by the Urban Redevelopment Authority (URA)[43] in 1976 to become the Central Area Open Space Plan. Parks and gardens were developed as areas of relief from the crowded city center, and new towns were planned with a host of hierarchically descending range of public spaces (Hee and Oi 2003). By incorporating land-use and transportation in accordance with the Singapore Masterplan,[44] the government planners opted for a new spatial and social order for public spaces in the city and in the new towns (Hee and Ooi 2003, pp. 88–89).

Singapore's 'Garden City' idea was attributed to the former Prime Minister Lee Kuan Yew who, in 1963, initiated the first Tree Planting Day to mark the beginning of the campaign for the greening of Singapore. There were important political and economic rationales behind the long-lived program. Due to the shortage of land for development and faced with the need to create massive alterations to the natural landscape due to the urbanization programs, the government had adopted a pragmatic environmental ideology. The early rationale for tree planting had been seen as a means of providing shade and shelter, trapping dust and reducing noise for public housing flat dwellers. In the 1960s and 70s, pragmatic concerns like the provision of shade and cooling of the environment were of topmost concern. The 'Clean and Green' city was also a boost for the thriving tourism industry.

In the three decades of the 'Garden City' campaign in Singapore, its emphasis had shifted somewhat from its original concerns. The success of greening the highly urban cityscape was such that by the 1990s, "Singapore (had) become a green, shady city filled with flowers, a city worthy of an industrious people whose quest for progress is matched by their appreciation of the beauty of nature".[45] There was a move to color the landscape and to create an aesthetic environment in the imaging of Singapore, seen as crucial to attracting foreign investment. Singapore was to exemplify efforts in taming the tropical jungle. Plants were introduced to cover

[42]The first Concept Plan of 1971 envisaged a Ring Plan, where the population was to be housed in satellite towns laid round the island—circumventing a green 'heart'—that would be connected by high capacity transportation routes, while the city center was to be hollowed out and redeveloped for commercial purposes. This move was to lay the foundation of the physical base for the economic development of Singapore.

[43]The conceptualization and implementation of the planning of Singapore fell under the auspices of the Urban Redevelopment Authority (URA), a statutory board initially instituted as the Urban Renewal Department (URD) in 1966, with immense powers in shaping the physical landscape—and until recently, answerable directly only to the government.

[44]The underlining basis of land use had been the concept of regulatory control through zoning, as crystallized in the Master Plan. This remains the basis of the Master Plans of 1965, 1970, 1975, 1980, 1985—right up to 1993, the latest Master Plan. To be read in relation to the Masterplan, the Concept Plan is an advisory and not statutory tool. It embodies the coordinating framework for public agencies and was flexible enough to accommodate changes in population growth trends.

[45]This was said by Lee Kuan Yew in 1980.

concrete structures like bridges and flyovers, while ornamental plants and shrubs graced roadsides and expressway planters. Instead of just wrapping the urban environment in foliage, it was necessary to cultivate nature as a 'playground' with easy access for residents.

Public parks and neighborhood gardens were created for the enjoyment of Singaporeans. The area allocated to parks and gardens covered 3 %, or a planning standard of 0.8 ha of parkland per 1000 of the population.[46] Although this figure is modest when compared with other major cities, it reflects the growing realization that as Singapore's population become more affluent, aspirations for a better quality of life would demand the provision of new recreational grounds and a pleasant green living environment.

This realisation has been matched by ambitious goals to 'bring people closer to nature'.[47] Singapore's Population White Paper, released in Singapore in 2013, outlines plans for at least 85 % of households to live within 400 metres of a park by 2030. These plans come with a paradigm shift in how the environment is perceived in the city. Singapore's plans to create a 'garden city' has now evolved into Singapore striving to become 'a city in a garden'—an augmented version of Ebenezer Howard's original ideal. To achieve this, Singapore has set greenspace targets and promoted community initiatives which allow people to have daily interactions with nature. The URA and NParks (National Parks) have pursued the concept of 'pervasive greenery,' a process of inserting greenery into every nook and cranny of a city's urban composition—from pavements and road dividers to vertical building facades and rooftops. Furthermore, urban gardening initiatives, such as the successful 'Community in Bloom' project, have also been encouraged as they help residents cultivate neighbourly bonds over the use of their common spaces.[48]

These green initiatives have been accompanied by 'blue' initiatives, which seek to 'bring people closer to water' by harnessing the potential of the water bodies on the island to provide clean waterways and aesthetically pleasing spaces of recreation for all (PUB 2014). In 2006, PUB (Singapore's National Water Agency) and URA launched the 'Active, Beautiful, Clean (ABC) Waters' master plan, which integrated drains, canals and reservoirs and redeveloped their surrounding physical landscapes. On top of performing their primary functions of drainage, flood control and water storage, these integrated waterways now serve as community spaces, as people can enjoy a more intimate connection with a resource that is vital to Singapore.

---

[46]This figure is outlined in the URA's 1991 Concept Plan.

[47]This is one of the principles in the Center for Liveable Cities' publication on '10 Principles for High-dense and Liveable Cities' (2013).

[48]The National Parks 'Community in Bloom Project' is a community gardening movement. There are now close to 1000 community gardens tended by over 20000 residents. See https://www.nparks.gov.sg/gardening/community-in-bloom-initiative.

2.2.1.4 New Landscapes of Public Spaces, Within Limits

With the economic boom of the 1980s boosting the growth of the city, there was on the one hand, an increasing interiorization as well as privatization of public space, as reflected in new typologies of spaces—the large air-conditioned atrium spaces of hotels and shopping centers, corporate plazas and atriums, as well as 'paid-for' public spaces like curbside cafes, shopping malls and cavernous discotheques. On the other hand, with the growth of public transport infrastructure, there was a proliferation of transit-type spaces, or "non-spaces", where one passed through instead of staying in, such as the foyers of mass rapid-transit stations, station platforms, bus interchanges, and pick-up or drop-off points for vehicles (Fig. 2.5).

As a player within the Castellian "global space of flows", Singapore played host to expatriates and guest workers who moved within international flows of labor and talent. Although foreigners have never been scarce in Singapore, the famed emporium of the East and the land of immigrants, the instability of the landscape of mobile people was of an unprecedented extent. The presence of groups of foreigners in public space on weekends was also a peculiar phenomenon of 1990s Singapore. These groups of guest workers form 'Sunday enclaves' which have a great impact on the public spaces in the city that they occupy. The attachment of these different groups with particular places and spaces in the city have become common knowledge—the Filipina domestic workers in and around Lucky Plaza on Orchard Road; the south Asian male workers in Little India, the Thai workers outside the Golden Mile Complex on Beach Road, and the Burmese workers around Peninsula Plaza on North Bridge Road.

2.2.1.5 The "Renaissance City"

With Singapore's rapid growth manifested in its built environment, the decade of the 1980s to 90s was a 'historical moment' of the arrival of the 'nation' in the economic sense. But the city became somewhat removed from its fleeting past, which some had perceived as a certain intangible lack of spirit and soul (Kong and

Fig. 2.5 Panorama shot of Citylink—a space of transit, "non-place"? *Source* Author's collection

Yeoh 2003, p. 132). Concurrent with the national drive to instill the neo-Confucian ethics embodied in the Shared Values for Singaporeans that guided citizens with the 'moral compass' drawn from traditional wisdom, the URA announced its Conservation Masterplan in 1989 to retain old city fabric of historical value. Attention was turned to the need to cultivate a 'softscape', with cultural areas centered on heritage buildings and historic quarters to create place-holders within the space of the global flow of people and ideas through Singapore (Ong 2004). The conservation of buildings and places also made sense for the growing tourism industry—visitors to Singapore would not be confronted with a global-city with no historical buildings and districts. The value of public spaces as venues for the appreciation of art and the performance arts was also emphasized in the Renaissance City Report of 2000—a long-term plan to unleash Singapore's art and cultural potential and to ultimately cultivate a gracious and cultured society (Fig. 2.6).

The location of the Padang in the area identified by the URA as the Civic and Cultural District in 1988 gave the space its new role as a site for culture. Events like the "New York Philharmonic Concert", "Merlion Week Concert", among others, took pride of place on the Padang. High profile recreational events like the Davis Cup tennis finals (1985) were broadcast to the world, with its backdrop the City Hall, Supreme Court and towers of the Central Business District—announcing to the world that Singapore had arrived in the global arena. It is not surprising that the Padang was featured as the 'poster space' of the URA's 2001 concept plan, with its

Fig. 2.6 Central Business District, featuring the Padang, 1982. *Source* URA Library Archives

Fig. 2.7 Panorama shot of Marina Bay. *Source* URA Library Archives

banner of "Towards a Thriving World-class City of the 21st Century". The backdrop of both historical buildings and skyscrapers created just the right mix of 'history', 'identity', and 'modernity' in a single package.

More recently, this 'poster space' distinction is arguably held by the Marina Bay area in Singapore, home to spectacular urban infrastructures such as the Esplanade Theatres, the Singapore Flyer (conceived by Kisho Kurokawa and DP Architects[49]), the Marina Bay Sands (designed by Moshe Safdie), Gardens by the Bay and the Marina Barrage. The Marina Bay area transformation was also part of the URA's Concept Plan of 2001, and it was realized in the later 2000s. The physical provision of public spaces also occupied an important place in the URA's agenda for the urban development, with its incorporation of its 2003 "Public Space and Urban Waterfront Masterplan"[50] within its strategies for improving the city center. The transformation of the Marina Bay waterfront includes mixed and multi-use developments, such as commercial, residential, and entertainment sites in the area as part of the URA's 'Live-Work-Play' vision. Marina Bay and its iconic features have raised Singapore's international profile as a hub of economic modernization, and positioned it as a tourist hotspot for years to come (Fig. 2.7).

2.2.1.6 A Short History of Speakers' Corner

As a response to popular pressure for some form of free speech, a Speakers' Corner—a la Hyde Park Corner in London, was launched on 1 September 2000, at Hong Lim Green, a small and nearly forgotten park in the city. Historically, the space at the edge of Chinatown was popular as a 'place for storytelling' during the Japanese Occupation of Singapore. In post-war days, members of the Singapore

[49]The Singapore Flyer is an attraction that was marketed to rival urban icons such as the Eiffel Tower and the London Eye. See Singapore Flyer website. http://www.singaporeflyer.com/about-us/design-concepts/.

[50]This was launched in conjunction with URA's new mission to "Make Singapore a Great City to Live, Work and Play"—an exhibition was held in June to August 2003 on "Our City Center—a Great Place to Live Work and Play!".

Fig. 2.8 Speakers' Corner (2016). *Source* Author's collection

Recreation Club and the Singapore Cricket Club used it as a cricket ground. But in the 1960s and 70s, it became popular again as a community space where Chinese operas were performed on a makeshift stage, and also for political rallies.[51] This tradition was recognized when it was designated the "Speakers' Corner". It was available for use on application for performance or public speaking and demonstrations, if the rules that regulate the space are complied with. Interesting events that have since taken place in the park included the annual Pink Dot SG event organized by the LGBT community of Singapore, as well as the arrest of young bloggers in September 2014 for staging an unapproved demonstration and disrupting a YMCA charity event at the park (Fig. 2.8).

2.2.1.7 At Home in Public

With the unqualified success of the public housing program, approximately eight out of ten Singaporeans now live in public housing, a hallmark of success in how public housing is now no longer perceived as low-income housing, but rather

[51]See the National Parks Board's website on Hong Lim Park. https://www.nparks.gov.sg/gardens-parks-and-nature/parks-and-nature-reserves/hong-lim-park.

homes for the national middle-class. The comfortable and roomy interiors of public housing flats were perhaps in part responsible for the low use of the space immediately around the flats. The popularity of TV and the Internet in Singapore meant also that to some extent, an abstracted and reduced concept of 'public space' could be brought to the home.

The public spaces that continue to attract the public are more 'concentrated' in its attractions. In new towns, such public spaces are found in the Town Center hubs and along busy nodes and routes, rather than in the quiet precinct public spaces. While new spaces in new towns become the public spaces of everyday life, overly prescribed spaces, such as the residential public spaces shared by eight blocks of flats that are called precinct spaces, are valued more as space markers and for their aesthetics than as actual lived space.[52]

2.2.1.8 Big Brother Is Watching, and Taking Your Temperature Too

Increasing levels of electronic surveillance in public spaces in Singapore, as in other cities in the world, from London to Los Angeles, have been justified under different agendas. These include occurrences of crime in popular public spaces like Boat Quay and Little India; cases of vandalism and littering, especially in the new town spaces such as corridors, lift lobbies, void decks and multi-storey carparks; and traffic congestion that is prevented with the Electronic Road Pricing System,[53] which work with both transmitters from devices installed in cars and video surveillance cameras at road gantries.

During the SARS epidemic of 2003, not only were thermal detection devices for body temperatures installed in public space, but individuals under strict home quarantine had surveillance cameras installed in their homes by the authorities as well (The Straits Times 2003a, b, c). Despite the high degree of surveillance, a survey conducted by the Straits Times in 2000 apparently found that most Singaporeans readily agree with the need for monitoring in public space: "It's like your big brother is watching you all the time. But if having a big brother means that I am safe from robbers and thieves, then I don't mind" (The Straits Times 2000). The privacy concerns were not considered an important issue, because, as one Member of Parliament explained, "you shouldn't be doing anything embarrassing in public" (ibid.).

With aspirations to become the world's first 'smart nation', Singapore has put forth the Smart Nation initiative to 'attract capital, talent, ideas', and to build an

[52]These are findings from a funded research project, "Design, Use and Social Significance of Public Space in Public Housing: Singapore and Hong Kong", National University of Singapore, for which I was the principal investigator.

[53]An extensive Electronic Road Pricing (ERP) system for monitoring road usage went into effect in 1998. The system collects information on an automobile's travel from smart cards plugged into transmitters in every car and in video surveillance cameras.

'outstanding urban environment'.[54] During his keynote address at the Smart Nation launch in November 2014, Prime Minister Lee Hsien Loong outlined his visions for Singapore to be a nation "where people live meaningful and fulfilled lives, enabled seamlessly by technology". He concluded his speech with the following remark:

> If I may go back to what Mr Lee Kuan Yew said when first we became independent and take it one step forward, update it, today perhaps this is what he would say: 50 years ago, we built a modern city. Today, we have a metropolis. 10 years from now, let's have a smart nation!

Singaporeans have already begun to embrace technology in their everyday lives, and these new advancements and abilities to harness big data can have many productive outcomes in enhancing public space. For example, many public events are now app-enabled, with users able to crowd-source real-time information to aid in decision-making, creating more efficient travel schedules. However, with the infusion of ubiquitous smart technologies into everyday life, issues of personal space and privacy become in need of protection, so boundaries between the public and private may need to be redrawn again.

2.2.1.9 Now in Public—The Breastfeeding Mothers of Singapore

In a way, the development of public space seemed to have come full circle. Within the framework of the developmental state's successful ascent to comfortable middle-class status, and in the absence of strife, political unrest or racial conflicts, one increasingly finds instance of public activism or even behavior bordering on civil protest in public spaces. Although not at the intensity of the post-war street protests and rioting, some of these events in public space certainly warrant one's attention and appreciation of the definite stirrings of organized civil society:

> A performance art of a different vein took place at the Esplanade Mall on Saturday. Onlookers gawked as about fifty mothers breastfed their babies and toddlers at Haagen-Dazs, an ice-cream cafe.
>
> Members of the Breastfeeding Mothers' Support Group and the Asia Parents E-Group, an online discussion group for parents, were gathered there to nurse their children. They were there in response to a recent incident at the Esplanade, when a woman breastfeeding her three-month-old daughter in the foyer was asked by a security guard to leave.
>
> The mum, Dr Hoe Wan Sin, wrote to The Straits Times. Her letter, published in the Forum page last Wednesday, prompted the two groups to organize yesterday's trip to the arts center.
>
> A spokesman for the Esplanade said: 'We're more than happy to have them here. The Esplanade is for everyone, including nursing mothers. We can empathize with their plight in public places, and we appreciate the difficulties they face'. It was an isolated incident, she said, and the security guard involved had since been briefed by the management (The Straits Times 2004a, b, c, d, e, f, g).

[54]These phrases were taken from the transcript of Prime Minister Lee Hsien Loong's speech at Smart Nation launch, Prime Minister's Office on 24th November 2014.

Calls for more public activism, though not yet cries of protest, are distinct and audible. These calls persuade authorities to view them as positive acts of civil society rather than as signs of dissent. Volunteerism and civic involvement are understood by Singaporeans and the polity alike, as "working quietly behind the scenes" within welfare or community organizations, instead of bold, organized acts in public space (The Straits Times 2004a, b, c, d, e, f, g).

2.2.2 Urban Trajectories

The mission of the URA, with the unveiling of the Concept Plan 2001, was "to make Singapore a great city to live, work and play…to build a vibrant, sustainable and cosmopolitan city of distinction". It has become apparent that the city-state needs to attract not just investment capital but "world-class" human capital to form the support base for the new "information society". The environment of strictly zoned environments decentralized throughout the island before the 1990s hardly inspired the 'vibrancy' sought after as the symbolic economy of a global city—the "street ballet" Jane Jacobs had lauded of New York City, or the bustle of downtown Tokyo—in contrast to the sterile business and industrial parks, lifeless business districts after dark and the ubiquitously bland environments of public housing estates in Singapore.

Clearly, Singapore has to re-imagine itself as a global city that would attract the international business elites or the jet-setting techno-preneurs to set up Singapore as their home-base for a while. Recognizing that the boundaries between business and services were blurring, the URA's key idea in the Concept Plan 2001 was to have a new zoning system in which industrial and business activities would be grouped according to "their impact on the surrounding environment". This new "impact-based" zoning approach allowed businesses to incorporate different usages under one-roof and change activities without resorting to the need for re-zoning applications, and was intended to create the potential for mixed-use environments in which to *work, live, learn and play.*

Within multicultural Singapore, public spaces are regarded by its politicians as important "hubs of activities, common grounds for forging community bonds, and a foundation upon which social cohesion and national identity can be built".[55] Public spaces are also seen as opportunities in the urban realm for imaging and creating Singapore as a global city, as exemplified by this vision:

> A stroll around the city centre will bring one to many parks and open spaces, linked by a pedestrian network of covered walkways, surface and underground links, and pedestrian malls along key activity belts such as Orchard Road and Singapore River. Streets of Chinatown, Little India, Kampong Glam and Bugis have become popular places where

[55]This was expressed by Vivian Balakrishnan, a Singaporean Minister, at a forum on "Civic Spaces in the Cities of the Asia Pacific" in March 2002.

Fig. 2.9 URA's PubliCity Event, 2015. *Source* URA Library Archives

> diverse cultures intermingle and communities congregate to celebrate. The Padang plays host to celebrations of Singapore's independence. Other attractive spaces such as Merlion Park, the "window" through UOB Building, One Fullerton's promenade, the atrium in Singapore Exchange, and Ngee Ann City Civic Plaza also enrich people's experience of the city (Skyline 2003).

One initiative by the URA is *PubliCity*. Part of the URA's Draft Master Plan of 2013, it seeks to activate public spaces in the city through good design, active programming and community engagement. The programme supports the concept of a "Street for People" and seeds community-initiated projects to transform streets into temporary public spaces.[56] Projects that have materialised include "Door to Door at Everton Park" (2015) where community members transformed a backlane in Everton Park into a temporary public alley and invited neighbours to join in a mini street festival (Fig. 2.9).

Plans are underway to enhance key activity areas in the city with a variety of public spaces through collaboration with private sector developments, the extension of linear green connectors within the city, an introduction of new urban waterfront parks with continuous promenades—plans that take into account aspects of programming, accessibility, climate, landscaping and public art. In September 2014,

[56]The URA makes a "Streets for People" how-to guide available for download on their website here: https://www.ura.gov.sg/MS/CarFreeZones/Events/Streets-for-People/~/media/CarFreeZone/Streets_For_People/PDF/Streets_For_People_How-to_Guide.ashx.

the URA invited members of the public to turn car parks into temporary public spaces, as part of the annual worldwide PARK(ing) event.[57]

These initiatives stem from the efforts of the state to capitalize on Singapore's status as a regional hub and global city and to enhance the city through new plans for integrated live-work-play communities with "vibrant street life". At the same time, these initiatives are locally grounded in the built heritage of the city. These are part of the state's strategy to attract and retain global talent in the face of global competition, as well as to retain its now more cosmopolitan and sophisticated citizens who could now choose to move to other cities. The new live-work-play environments espouse overlapping networks of spaces related to these different functions and networks of people. They create new opportunities for 'meeting' in the city and encourage a more cosmopolitan character in the city's public spaces.

[57]PARK(ing) Day is an annual event where people transform parking spaces into temporary public parks. See the URA's press release: https://www.ura.gov.sg/uol/media-room/news/2014/aug/pr14-47.aspx.

Chapter 3
Case Studies

3.1 Selection of Case Studies

The social, cultural and political milieu that forms the backdrop of Singapore public space is a singularity, as it is shaped by the unique circumstance of Singapore's history, geography, people, as well as the city-state's engagement with aspects of colonialism, independence, nation building and globalization. Singapore's urban and spatial configuration had been formed by directed developments in its relatively short history, these being in particular:

- the adoption of the Ring Plan in the 60s
- the implementation of land-use zoning in the 70s
- the exodus of the population from the central area to new towns at the periphery from the 70s
- conservation of historic ethnic districts in the 80s
- globalization and the constant re-imaging of the city from the 90s and 2000s to the present.

Until recently, nation building had been less of a historically continuous process and more constitutive of perceptible phases of development, concretized in the built (and unbuilt)[1] history of the city. The first time visitor to the city of Singapore is wont to have such an impression:

> Indeed, to walk about on the Singaporean streets and squares makes one immediately aware that in this city quite different layers of urban sensation converge. The architecturally impressive skyline in a decidedly modernist style and the elegant shopping malls on Orchard Road, on the one hand, coexist with the labyrinthian network of streets and by-streets… of Little India, Chinatown and the Arab Street on the other hand…the urban design (sic) brings together completely divergent strands of contemporary urban culture (Paetzold 2004).

[1]Here denoting the planned histories, demolished environments and remembered places.

L. Hee, *Constructing Singapore Public Space*, Advances in 21st Century Human Settlements, DOI 10.1007/978-981-10-2387-3_3

In the imagery of Singapore on postcards and in the multitude of tourist guidebooks and brochures, the ubiquitous pictures are the ones of the central business district skyline and the international marketplace of Orchard Road. In Singapore, the notion of "the city" evokes the cosmopolitan culture of the global city, which "manifests the global citizenship of Singapore much more powerfully than any sense of the local and the historic" (Bishop et al. 2004, p. 26). Sites like Orchard Road in particular embody the transnational orientation of the city, where shops selling international labels mark the street as a portal to international culture, fashion and commodities, and where citizens from far-flung lands rub shoulders with well-heeled Singaporeans.

Conservation efforts of the 80s and 90s have also left the city with the built legacy of the shophouse districts of Chinatown, Little India, Kampong Glam and Tanjong Pagar, among others. These were the historic cultural districts, where different ethnic groups had thrived during the British colonial rule. Within these ethnic quarters, one finds varying degrees of gentrification or surprising glimpses of intact ethnic communities, depending on where one looks. These ethnic quarters are often depicted in tourist brochures as definitive sites of Singaporean cultural life. Among these districts, Little India embodies a good mix of ethnic communities, a continual influx of foreigners from the Indian subcontinent, and the impacts of modernization.

In contrast to the more cosmopolitan and cultural image-spaces of Singapore, in nation building pamphlets, posters or advertisements, or where images of the notion of community are depicted, Housing and Development Board (HDB) flats from the Singaporean "heartland" form the quintessential backdrop.[2] Indeed,

> (t)he representations of the heartland…paint a picture of the city-state outside of the city, rejecting its skyline of tall corporate buildings for that of uniform HDB blocks. The divide between the city and heartland is made of certain erasures and elisions (Bishop et al. 2004, p. 26).

The drastically different landscapes of the new towns are borne out of the tabula rasa destruction of extant topographies, built environments or settlements. They embody the image of relentless progress, upon which the national polity wages its political legitimacy. As sites for the 'building' of a national identity, the new town landscape is so unique that it could be read as exclusionary. Here, we depart from the notions of the global city into the confines of a distinctively Singaporean vernacular. The HDB estates form the physical milieu of home and community life in Singapore.

From the diverse spatial forms of urban Singapore, I have selected the three important sites of Orchard Road, Little India and HDB public housing spaces as the

[2]The term "heartland" is used synonymously with the HDB new towns, as an idiomatic term for the collective majority of the population. The term is often used to evoke antagonism between "heartlanders" and "cosmopolitans", emphasizing the divide between the rooted communities in new towns and the mobile, highly educated elites who are citizens of the international community and are not confined to the housing estates.

contexts to investigate the constructions of public space. It is important to explore a range of diverse public spaces, and the sites selected here encompass many different orientations and functions of public space, as well as embody diverse spatial practices and meanings. These, by no means, represent the whole range of Singaporean public space. My intention is not to represent its totality, but to represent its different logics, which, I believe, will yield rich findings.

3.2 Orchard Road as Conduit: Between Nostalgia and Authenticity

> Orchard Road is the East's answer to New York's famed Fifth Avenue, Paris's Champs Elysee, London's Mayfair and Sloane Square and New Delhi's Janpath, all rolled into one. The only competition it may face in the very near future could be from Nanjing Road in Shanghai or the ultra modern and fast and dazzling developing eastern end of Jianguo Avenue in busy bustling Beijing (Abraham 2003).

Singapore's Orchard Road is a 2.4 km long street characterized by the Singaporean's love for mobility, mass consumption and the occasional spectacle of mass recreation. At first sight, it seems to be yet another glitzy shopping street of a cosmopolitan city:

> Orchard Road is the prototype for a new shopping typology that is becoming more prevalent in Asia: the urban linear mall. This type of mall rescales the plan of the traditional mall…to one that amasses the spaces of the city (Vinh 2001, p. 208).

To make a case for understanding Orchard Road beyond the popular but essentialist regard of it as a linear shopping mall, I will frame the discussion in terms of its relevance to Singaporeans and Singapore urbanism as an authentic urban public space.[3] As a 'post-traditional' space, Orchard Road is a portal to new identities being constructed of place, as well as of the inhabitants of its space, its past being at the same time selectively re-imagined and re-constructed. The street has undergone vast transformations since the 1800s, shifting through narratives of space, with networks of places, times and bodies supplanting one another. The teeming ethnoscape (Appadurai 1996), comprising tourists and transnational workers of various social strata, is highly visible at the main street of the nation. Actors in global epistemic communities, globe-trotting *flâneurs*, urban nomads, the middle-class Singaporean consumer, and workers of the 'domestic economy'—female domestic guest workers—share the public spaces collectively strung together as Orchard Road. Orchard Road is the quintessential public space of the city, where different groups of people are channeled into close proximity—sometimes the 'space of appearances' and at times the 'space of friction'.

---

[3]Parts of this chapter have been published as "Singapore's Orchard Road as conduit: Between nostalgia and authenticity" in *Traditional Dwellings and Settlements Review* 17, no. 1 (2005), pp. 51–63.

It is difficult to make a direct comparison of Orchard Road to other famous shopping streets of the world, as each is defined by its historical and urban context. It is useful, though, to think about Orchard Road alongside some other models, as this reveals what it is not and helps us to define what it is. It is not a linear urban mall like CityWalk at Universal City, California,[4] which is a wholly conceived shopping street-based mall, a sort of retail-based connective tissue predicated on consumption-induced urbanity. Like "brick-street" Ginza of the nineteenth century in Tokyo, which was intentionally located near foreign residents' settlements, Orchard Road was developed as a tourist-oriented shopping street due to its proximity to expatriate communities. And like pre-war Ginza, which was a hangout of fashionable *mogas* and *mobos* (modern girls and boys), Orchard lures its youthful crowds with the latest fashions, movies, discotheques, cafès, and food from around the world. However, Orchard Road is not comparable with the scale and complexity of the Ginza district of today, with its large-scale interior environments of *depātos*.[5]

Of a more similar scale would be the *Kurfürstendamm* (or *Ku'damm*) of Berlin, which stretches for 3.5 km and was built in the late nineteenth century to emulate, albeit on a more modest scale, the *Champs-Elysées* of Paris. Like the modern international showcase of the *Ku'damm*, once touted as "das größte Caféhaus Europas", Orchard Road boasts fine hotels, departmental stores, restaurants, cinemas and art galleries. It is similarly served by three subway stations, and its street scene thrives despite the heavy traffic on the street. Like the *Ku'damm*, Orchard Road is sometimes closed to traffic and transformed for carnivals, festivals and parades. However, unlike the *Ku'damm*, Orchard was not conceived and rebuilt as a coherent shopping and entertainment street,[6] but grew piecemeal over time, shifting through different characters and times. Orchard Road is not a store-fronted street, but consists mainly of disparate buildings with deep interiors, joined by the pedestrian mall, which, interestingly, pre-existed in anticipation of the buildings that had yet to materialize when it was built in the 1970s.

The lack of uniformity on the Orchard Road frontage gives the street a different character from many European shopping streets. It engenders a secondary economy of street-side kiosks and an informal economy of vendors of various goods, from ice cream to on-the-spot caricatures. In terms of urban public spaces, Orchard Road presents a wide range of sectional profiles and types of spaces of different scales and relationships with the built structures on the street (Fig. 3.1).[7] The uniqueness of Orchard Road stems from the plethora of activities one can engage in, besides shopping. These activities are not limited to the insides of malls. There are countless pushcarts, kiosks, mobile food trucks, temporary sidewalk stalls,

[4]Universal CityWalk was designed by Jon Jerde in 1992.

[5]Depatos is the Japanese term for departmental stores.

[6]The Ku'damm was to a large extent destroyed in WWII, and many buildings were rebuilt. However, the framework of the street set up by Otto von Bismarck in 1875 is still discernable.

[7]The different sectional profiles found on Orchard Road share adjacencies that are difficult to represent on an urban scale plan.

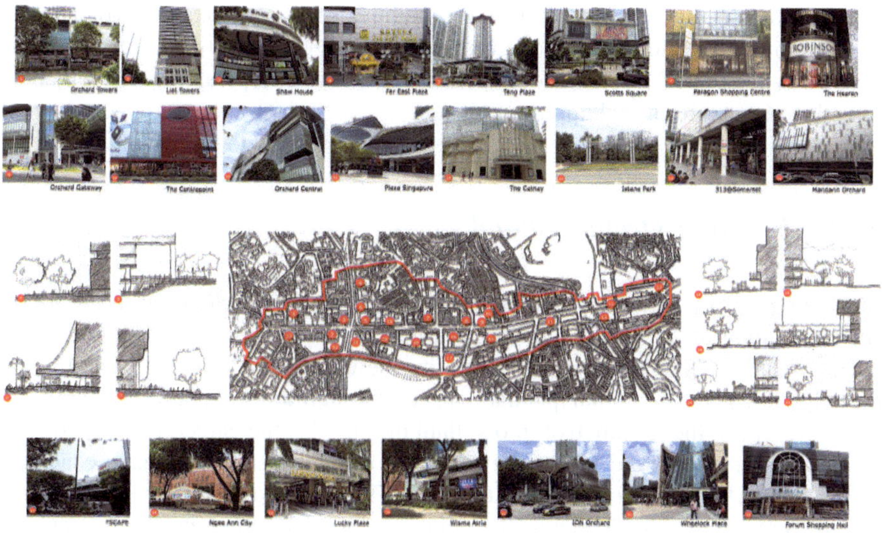

Fig. 3.1 Orchard Road landmarks

product-promotion marquees and even temporary stages for street performances, as well as a whole range of interiors, in-between, and exterior environments that one can inhabit.

Materially, the post-traditional moment of Orchard Road is one in which its architectural substance shifts from solid building forms and shapes (as embodied by the old Mandarin Hotel and C.K. Tang Building), to new typologies predicated on surfaces, such as video-walls, changing seasonal facades, and even morphing walkway paving patterns in a constant updating of images.

The new ephemeralism and instability of appearances are concomitant with the displacement of bodies—the changing clientele of global *flâneurs*, the dynamic ethnoscape and the shifts in affinities to spaces. Sometimes, this occurs not out of choice.[8] Even the forms of representation of Orchard Road has changed from traditional figure-ground maps and photographs, to abstract forms like graphs anticipating real-estate and rental values (peaking around the Scotts-Orchard intersection and falling in the middle), demographic patterns (age-groups of consumers, shopping preferences, petty crime statistics), and scopes of surveillance (camera-surveyed atriums, MRT stations and sidewalk bank automats; the prominent Electronic Road-Pricing (ERP) gantry).

The verdant green of the erstwhile tropical landscape remerges now and then in startling bits, not least in the green fringe of the stately angsana trees reminding one again of the one-degree north-of-the-Equator location of this shopping street. Pools

[8]For example, the opening of Ion Orchard in 2013 which entailed the eradication of the open green space above the Orchard MRT station led to the displacement of Sunday enclaves of Filipina domestic workers.

of tranquility in the greenery around the vicinity of the lower stretch of Orchard Road espouse languid tropical calm, a real change of the urban nature from the more built-up areas. And it rains on Orchard Road—real tropical torrents, which are difficult to re-create in a shopping mall.

3.2.1 Framing Orchard Road

The analogy of the actual geography of Orchard Road as a cleavage in the topography—an indent—is a useful concept to illustrate its metamorphosis through time.

Historically, Orchard Road had begun its useful life as a kind of trough that ran through the hills of colonial plantations, a thoroughfare navigating through and collecting from the productive plots flanking it. As the street developed as a vehicular artery, it served too as the space for a much needed drainage canal that alleviated the flooding problems of the area. The Orchard Road of the 1960s and 70s Singapore would have seen the beginnings of a more pronounced topography—the indentation of the channel was deepened as the buildings flanking it take over the place of gentle hills. Orchard Road served its new role of concentrating and distributing commerce and consumers along its lucrative stretch. By the 1980s and 90s, Orchard Road already reflected the honed tastes of the Singaporean consumer for endless choices of high-end goods presented in sophisticated environments, the love for speed (fast food, transit), as well as a caffeinated lifestyle stretching well into the bright lights of the night. Orchard Road has become the conduit transmitting the energy and sped-up pulse of the city, linking its networks of air-conditioned atria and open public spaces both above ground and through the subterranean mass rapid transit system.

The conglomeration of these spaces which we call Orchard Road is certainly more than the sum of its parts—when I refer to Orchard Road as public space, it is the concept of the street, its physical and embedded structures, visible and invisible, as well as its narrative through time that are referred. Through the reference frame of a 'flowing channel' (both literal and figurative), I situate Orchard Road as the juxtaposition of elements which are ultimately emblematic of the post-traditional environment of the city-state: colonial history, mass consumption, the symbolic economy, global workers, tourism, the Singaporean middle-class, greenery, local history and spatial practices. Does the geography of Orchard Road offer us a connection with its history? The construction of Orchard Road as the public space of the city through the montage of these erstwhile or embedded elements provides us with insights to the post-global moment of Singapore. In particular, the following spatial moments will be explored in this chapter:

1. The collision of history, values, contemporary culture and the symbolic economy—how are these represented or embodied in the present?
2. The divided space of global and local actors, as well as the formal and informal economy—how do these different groups come together in space?

3. The disjuncture of fragmented personal experience, memory and urban phenomena, through which concepts of public space are formed—what is the nature of nostalgia for the post-traditional environment of Orchard Road, and what is authenticity in this regard?

Much like how Walter Benjamin had perceived the trajectories presented by the elements of the arcades of nineteenth century Paris (Buck-Morss 1991), Orchard Road is a projection of the future of urban Singapore—allowing its disparate elements to reassemble, and at the same time, to relate with each other. These elements act like hypertexts, projecting into whole realms of possible futures and discourses. In the case of Orchard Road, these would include: shopping, mass transport, tourism, entertainment, fast food, globalization, urban renewal, state control, youth culture, and real estate values. Orchard Road acts like a portal into the future, concentrating and connecting otherwise disparate urban discourses within the constructed discourse of urban public space.

The post-traditional condition of Orchard Road, as offered by these constructions, is not based on a history of progress or even a progression of history, but rather on the compressed moments of time, images, places, people and scales of urban phenomena.

3.2.2 Orchard Road—A Brief History

Early settlers of Singapore had originally cultivated the hilly area as gambier plantations,[9] but these were given over to colonial merchants for the cultivation of cash crops, principally nutmegs, by the 1840s. The hitherto unnamed road that ran through the plantations appeared in the Plan of Singapore Town of 1846 by J.T. Thomson, the first Government Surveyor, as Orchard Road. The hills flanking the road had been given the names of the estate owners, such as Oxley Rise, Claymore Hill, Cairnhill Hill, Mount Sophia, Mount Elizabeth and so forth. Unfortunately, disaster struck in the 1850s, when an unknown crop disease killed the mostly monocultured cash crop, and this ended the agricultural ambitions of Orchard Road. The plantations were sold and divided into smaller residential plots and Orchard Road became an area of airy colonial bungalows (Fig. 3.2).

By the 1880s, Koek's Market and the first shophouses had been built, although the nature of the street was still very much rural. The shops found here were utilitarian—sundry goods stores, and later, motor vehicle workshops. The old Orchard Market was pulled down in the 1920s, and the area continued to be flood-prone due to the topography of the street—an indent between hills—until the building of the Stamford Canal in the 1960s. While horse-drawn carriages and *jinrickshas* were the common modes of transport in the 1920s, these changed

[9]Gambier was used for tanning and as a natural red dye, but it was also a component of betel nut chewing, a favorite pastime of field workers, back then.

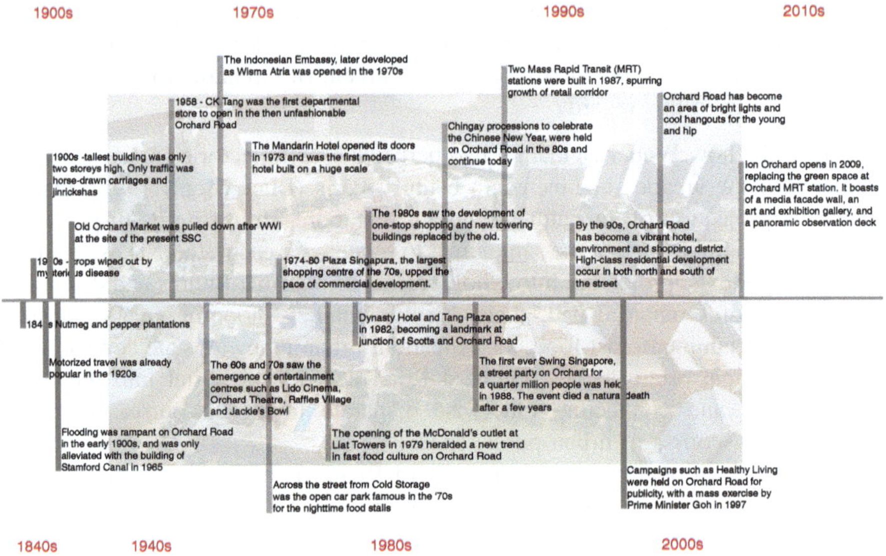

Fig. 3.2 Orchard Road timeline

dramatically by the 1940s to busy automobile traffic. By then, Orchard Road was beginning to take on the mantel of a small scale shopping street with the addition of more shops and three supermarkets in the 1950s. A milestone was reached when in 1958, C.K. Tang, an enterprising local merchant, set up the first departmental store there when it was still unfashionable and largely undeveloped. It was not till the 1970s that Orchard Road replaced High Street as the main shopping street of Singapore.

The 1960s and 70s also saw the development of Orchard Road as a local entertainment strip, with the opening of Lido Cinema, Orchard Theater, Raffles Village and Jackie's Bowl. Across the street from Cold Storage Supermarket, an open carpark by day served as a popular site for open-air food stalls by night. The Mandarin Hotel, opened in 1973, was the first large-scale modern hotel to open here. With the 1974 opening of Plaza Singapura, the largest one-stop shopping center at that time, the pace of commercial development on Orchard Road stepped up. The Orchard Road Pedestrian Walkways,[10] a state-implemented project which had begun in 1976, provided the wide paved sidewalks flanking the street, adding to the imagery and aesthetics of an upscale shopping street. In 1979, the first

[10]The provision of walkways was completed in two phases—in 1976, under the Public Works Department's "Walkway Program" and in 1989, under its "Ten Year Walkway Masterplan". The second program involved the aesthetization of Orchard Road with the installation of decorative street lamps, railings, seats, "pocket parks" and embedded brass motifs of tropical fruits (the "orchard").

MacDonald's outlet opened its doors in Singapore, at the junction of Scotts Road and Orchard Road, heralding the debut of the new phenomenon of fast food culture.

Significant developments in the 1980s were to change the face of Orchard Road, such as the building of the iconic, if somewhat eclectic, Dynasty Hotel (now Marriott) at the prime junction with Scotts Road, and the opening of Tang Plaza at its podium (replacing the old landmark of the C.K. Tang building). Other towering buildings replaced many of the two-storey shophouses on the street. The opening of three mass rapid transit (MRT) stations in 1987 spurred further growth of the retail corridor, which was anticipated to stretch all the way to the new Marina City in the south. By the 1990s, Orchard Road had become a busy hotel, entertainment and shopping street, with high-class residential development in the northern and southern areas. Large, self-contained malls like Ngee Ann City offered not only a myriad of shops, restaurants and departmental stores, but also a whole host of public functions including a public library, post office and a "civic plaza". At the turn of the 21st century, Orchard Road had all the hallmarks (and brand labels) of an international shopping street—a new image of bright lights, large screen video walls and a cosmopolitan crowd sitting at its numerous al-fresco curbside cafés.

Spectacular events such as the colorful Chingay Procession during Chinese New Year were held on Orchard Road in the 1980s. The first ever large-scale street party, Swing Singapore was held in 1988 and saw a turnout of about a quarter million people.[11] The annual Christmas light-up of Orchard Road also began in the 80s, during which the entire street was transformed by colored lights (and a lighting consultant or two) into a simulated Winter Wonderland. Ad hoc events, like the Healthy Lifestyle Campaign of 1997, with the then Prime Minister Goh Chok Tong himself leading exercises for the masses, also took place on the main street of the nation. The high visibility afforded by Orchard Road as an event space culminated in the Millennium Countdown Party which was televised worldwide and featured the public space of Orchard Road in the same take as Tiananmen Square Beijing and Times Square New York.

> Rococo pagodas perch atop slippery-flanked megastructures concealing enough cubic footage of atria to make up a couple of good-sized Lagrangian-5 colonies. Along Orchard Road, the Fifth Avenue of Southeast Asia, chocka-block with multi-level shopping centers, a burgeoning middle class shops ceaselessly. Young, for the most part, and clad in computer-weathered cottons from the local Gap clone, they're a handsome populace; they look good in their shorts and Reeboks and Matsuda shades (Gibson 1994).

The youth market on Orchard Road is definitely a growing niche, with specialty buildings in the 1990s such as The Heeren and Cathay Cineleisure Orchard catering primarily to young consumers, and the Youth Park and Skate Park adjacent to the National Youth Center catering to their recreational and social needs. Another demographic group changed the constitution of the Orchard Road crowd, at least on

[11]Street closures for street parties continued on a monthly basis, but the practice fizzled out after a few years. The novelty of street parties had apparently worn out.

Sundays—the large groups of female domestic workers from the Philippines, many of whom spent their day off hanging out at particular spots on Orchard Road.

In 2003, 4.56 million visitors or 76 % of all visitors to Singapore visited Orchard Road.[12] Although the globetrotting tourist can be found in many places on Orchard Road, a closer look at their presence in public spaces would establish that in a very general sense, white expatriates tend to gravitate towards Emerald Hill, at the Peranakan Place's perpetually carnival-like atmosphere, while the Asian tourists tend towards the Tanglin end of Orchard, near the large duty-free malls. To many expatriates, Orchard Road is in fact home, as their expatriate contracts would quite often furnish them with an apartment with a view of the busy street.

The Orchard Road of today is certainly a far cry from the plantation days of the colonial period or its early days as a local commercial street. The global outlook and cosmopolitan orientations of its clientele bring on a concentrated edge situation where differences become apparent and one becomes acutely aware of the elements that bring the city into the global age. Orchard Road has even received international acclaim, with a 2012 report by Presence Mystery Shopping, a Paris-based market consulting company, ranking Singapore as the world's best shopping street (CNN Travel 2012). We will now look more closely through the lenses of analysis offered in this chapter at the kinds of spatial moments that animate Orchard Road's post-traditional public space.

3.2.3 Compression of Time and Images

Orchard Road is a street that grew up over time, and it still embodies large parts of its history and topography within its urban fabric. Indeed, the historical nature of Orchard Road still determines to a large part the nuances which color the archipelago of public spaces at different stretches of the street. These are nodes of activity floating in the sea of urban matter, a reversal of the traditional urban figure-ground gestalt, joined by the currents of movement in-between. The following local "districts" can be identified along Orchard Road.

3.2.3.1 Districts

The Orchard Towers (1975) sits as a relic of the 1970s along the stretch of new or refurbished hotels and upscale malls. The public space just outside of Orchard Towers attracts by day a motley crowd of foreign workers who visit the employment agencies within the building, and by night and on weekends, crowds of

[12]...and spent S$340 million, which is 15 % of the total shopping expenditure by visitors in Singapore, according to Minister Balakrishnan, in his speech at a press conference for "Sharing the Vision for Orchard Road's Evolution", March 2005.

foreigners, especially sailors from the ships that berth in the busy Singapore harbors, who frequent the euphemistically named escort agencies and bars within the building. Orchard Towers, in fact, is a so-called "Designated Red-Light Area" (DRA) in Singapore, meaning that the building is legalized as a site of prostitution. However, nothing on the public facade of the building suggests that fact, except for the type of crowd seen on this particular stretch. A little further down the street, the Thai embassy sits within its own fenced compound, surrounded by serene greenery.

The prime real estate location of Orchard Road is surely the junction of Scotts Road and Orchard Road. This junction is punctuated by large commercial buildings, namely the eclectic Tang Plaza and Marriott Hotel (1982), Shaw House (1993) and Wheelock Place (1994). Here, major brand name labels are literally festooned on the facades of buildings, and a large video wall plays on the Shaw House façade. This is where buildings reveal their insides on the outside. It is the place to be seen on Orchard Road, sipping coffee at Starbucks or just hanging out further down the street at the Ngee Ann City (1993) Civic Plaza fountain. This is also one of the busiest pedestrian nodes of Orchard Road, as the Orchard MRT Station (1987) spews out its payload on the fourth corner of this junction. Lush, old angsana trees still provide delicious shade on the pedestrian mall. The public spaces at the Ngee Ann Civic Plaza and at Shaw House are also the favorite spots for the white marquees set out on weekends for product promotions or public events like Chinese Lion Dance performances.

The next nodal point on Orchard Road is the domain of the young and trendy. The young crowd spills out from the underground Somerset MRT Station to Somerset@313 and a little down Grange Road, Cathay Cineleisure Orchard.[13] These lifestyle malls are targeted at capturing the attention of the young shopper, boasting popular dessert outlets, and annexes dedicated to youth fashion. Right across the street from Cathay Cineleisure Orchard is the Youth Skate Park, which sees a constant stream of teenagers at different times of the day and week. Adjacent to Cineleisure Orchard is *SCAPE building which attracts dance enthusiasts who practice their latest moves in front of the studio mirrors lining the plaza, musicheads who come in throngs to events such as the YouTube FanFest Singapore held at *SCAPE in 2015, and secondary school students who attend entrepreneurship workshops at the youth foundations who have set up their offices there. Overall, the youthscape of Orchard is a temporal state that comes alive particularly on Friday nights.

The pastel-colored facades of Peranakan Place, a conserved row of late nineteenth century shop-houses of the Straits-Chinese style, mark the entrance to Emerald Hill, also a conserved district of shop-houses off Orchard Road.[14] The area was originally slated as a showcase of Straits-Chinese culture and lifestyle, but it

---

[13]This youth landscape was once also defined by the Heeren from the late 1990s to the late 2000s, especially with its anchor tenant, record store chain HMV.

[14]Peranakan Place and Emerald Hill Projects were conserved by the URA and assigned new uses in 1984.

suffered from a lack of critical ‘cultural mass’ to attract visitors. Instead, the area has found new life as an enclave for beer-drinking tourists and expatriates, with the alleyway offering Singaporeans a showcase of expatriate hang-outs and pick-up joints.

The Istana Park and Istana Gate further down Orchard Road take on a completely different character. The Istana building and grounds had been here since 1869, when it was completed by the British colonial government as its headquarters, then known as Government House. Here, the iconographic space is not so subtly emblazoned with motifs of statehood in general and of the Singapore flag in particular. This part of Orchard Road embodies the tranquility that comes with matured trees, pools of reflecting waters and sleepy pavilions in the Istana Park. On national holidays, crowds of Singaporeans who do not normally shop on Orchard Road would form queues to get into the Istana Gardens for a rare visit.

3.2.3.2 Plans

The vintages of the buildings on Orchard Road hint of its asynchronous development—embedded within the area are many different plans and visions, such as the colonial masterplan of Singapore, the tourist-oriented developments of the 1950s and 60s, the URA Development Guide Plan for Orchard of 1994 and the Singapore Tourist Board (STB) Masterplan of 2001 for “Making Orchard Road More Happening”.[15] The 2001 plan by the STB shows a clear move towards having more dialogue with private sector interest groups,[16] such as the Orchard Road Business Association, and also in giving planning incentives, such as additional allowed gross floor area (GFA) for “activity-generating uses or outdoor refreshment areas on the roof decks of ground links, in second and third storey links, underground links and on urban verandahs” (Seow 2001, p. 10) Orchard Road had hardly developed in a ‘natural’ fashion, but neither had it developed exactly according to the blueprint. A close look at many of the planned developments over time reveals that ideas have been discarded, modified or augmented in some ways to deal with the economic reality of development or the level of viability of the plan for those who have to implement it.

Orchard Road, far from being a homogenous manufactured reality, has been constantly repositioned to meet the challenges posed by the particular moment in time. The legitimacy of Orchard Road’s history lies not so much in its built history, but really in its ‘geography’—that it had grown always in the form of a channel or

[15]In 2005, URA launched a series of initiatives to “enhance and rejuvenate” Orchard Road. These involved the setting up of an Orchard Road Development Commission to review redevelopment proposals and relaxing various architectural and urban design-related guidelines, with the aim of making Orchard Road “a great street”. https://www.ura.gov.sg/uol/media-room/news/2005/mar/pr05-10.aspx.

[16]It is interesting to note that after the public exhibition and consultations with various public and private agencies regarding the proposals, some ideas were considered impractical and dropped.

conduit over time, and through different urban ideologies and moments. The street is the linear narrative that strings together the temporally disjunctured pieces of Orchard Road and conflates these images, at the same time projecting these into the present. The forces driving Orchard Road's development—history, planning agendas, real estate values, contemporary culture, changing tastes and habits, and the political will to create the appropriate imagery of a 'global city of the 21st century'—sometimes work synergistically but often antagonistically, so that the end product is a negotiated outcome rather than a planned one.

3.2.3.3 Times

Orchard Road is very much a place that has its own temporal cycles of events. These occur at various time scales: on daily, weekly, seasonally and yearly cycles, and include both 'staged' and ad hoc events, and those occurring out of spatial practices of the various groups of people inhabiting the public space of Orchard Road. While the early mornings on Orchard Road belong strictly to the road sweepers and kiosk vendors, Friday nights transform the street into the favorite hangout for the young and trendy; Saturday afternoons witness the barrage of militant shoppers, some pushing baby carriages like they were tanks! Orchard Road during the SARS outbreak of the spring of 2003 proved to be the most visible barometer of the national health scare—the street was painfully deserted even on the best shopping days. The Great Singapore Sale, now just past its twenty-second anniversary, entails a whole plethora of public events, including a fashion catwalk held on the street.

3.2.3.4 Images of Elsewhere

The desire by the government, through its agencies such as the Singapore Tourist Board, to bring Orchard Road onto the world stage—in a coming-of-age gesture—might be witnessed in its enthusiasm in the 90s to hold "city twinning" events, such as that of Orchard Road with the Ginza, Tokyo, or Orchard Road with the *Champs-Elysées*, Paris. Weeklong festivities such as fashion shows, tea ceremonies or wine festivals bring the Ginza and *Champs-Elysées* to Orchard, while Orchard Road is at the same time mirrored in Ginza and the *Champs-Elysées*. Orchard Road is the conduit that beams Singapore across the world.

The ultimate simulation of time and images must be the hyperreal re-creation of a Winter Wonderland replete with snow-laden buildings, icicles and reindeer sleighs every year, for about two months over the Christmas and New Year season. The Christmas light-up (Fig. 3.3), after 20 years, has become a "new tradition" or even a simulacra of a Christmas tradition in Singapore, although only 19 % of Singaporeans are Christians who actually celebrate Christmas as a religious festival[17]:

[17]From 2015 Census data published by Singapore Statistics.

Fig. 3.3 Christmas at Orchard Road

(From "Jonie", Eurasian girl, X'mas 2001)

> Being of mixed parentage, I have grown up with a very traditional Christmas which is celebrated in exactly the same way year after year. For us, the feel of Christmas comes once the first sight of decorations hits Orchard Road (the shopping hub of Singapore) and it is usually in early November as the lights are put up to get everybody in the mood (and promote shopping!!). Everyone busies themselves(sic) with personal shopping, new shoes and clothes for the kids, plus gift-buying.[18]

Phenomena like snow, the Northern Lights, and traditional symbols of Christmas and the winter season—like fir trees, angels, bells and snowmen, are writ large—on the urban scale of the street, to the degree of becoming hyperreal. The transformation of space creates temporary new nodes and activities, such as the experience of snow at Tanglin Mall. Contributions to the public realm in the form of decorated facades and public spaces by private enterprises, in return for increased people traffic, blur the lines between the public and private. There is also a (temporary) suspension of difference within the transformed environment—due to the increased crowdedness of the street, niches occupied by different groups become less obvious. But because the Orchard Road Christmas light-up has worked its way into both collective and personal narratives, it is sustained as a public event, turning Orchard Road into the public space of the national collective during the holiday season. The streets are crowded but the crowds tolerant—there is a real sense of wonder at the

[18]This excerpt was taken from a website that is no longer active.

transformation of the street. With fervent photograph-taking by groups, friends and families—the notions of 'event' and 'festivity' are real:

(From an Internet entry by "Y.K. Chia", father of two, X'mas 2001.)

> Orchard was packed with merry-making crowds. The rain had stopped, it was cool and pleasant. This year's deco theme was "Northern Light" - holographic paper on tubes (more than 50,000) were stringed across the streets creating rainbow- hue 'curtains'. Not quite the real thing but a welcoming break from the 'false' snow and reindeer theme. Makeshift stores selling blinking magic wand, light bars, girls with plastic 'wings' going around asking for donations. Crowds gathered to watch daring young boy yo-yoing up and down secured with elastic band on the catapult- like Y structure. We board the North bound MRT and headed for home. That night the two slept with their new 'play-thing'- light loops beside their beds. Xmas arrived in their dreamland.[19]

3.2.4 Compression of People and Practices

Every Sunday, the shopping center Lucky Plaza is transformed into a site where temporary enclaves of Filipina domestic workers congregate. These women collectively spend their weekly day off in the shopping center, their favorite rendezvous spot right in the heart of Orchard Road. These Sunday enclave exhibits spatial practices not typical of Singaporeans in general. This was especially the case with the picnicing and group socializing that took place in the green space behind Orchard MRT station, which has now been taken over by Ion Orchard. These acts of spatial appropriation, along with loud chattering and laughter, are viewed with some degree of ambivalence by Singaporean shoppers and with some amazement by tourists on the street. Occasional brushes with the building management about their right to hang-out at Lucky Plaza have occurred in the past, but the Filipinas have since re-claimed the atrium space as a rendezvous point. The following excerpts from "Sunday blues for loiterers: Maids not welcome at Lucky Plaza" (Vasoo 1998) sum up the different points of views:

> Loud chit-chat among them also affected business and transactions, with shopfronts and doorways blocked by throngs of loiterers.
>
> Lucky Plaza Management Committee[20]

> I don't understand why they are driving away our best customers. Sunday is the only day where(sic) the maids are allowed to come out and the locals don't come on that day, so why do they have to set up such a policy?
>
> Lucky Plaza shopowner

> Now I just go in and remit my money. I don't want to spend too much time there.
>
> Ester Santiago, Filipino domestic worker

---

[19]This excerpt was taken from a website that is no longer active.

[20]...on its measures to stop domestic maids from loitering around the shopping center on Sundays.

3.2.4.1 Sunday Enclaves

The phenomenon of these Sunday enclaves (Fig. 3.4) is one manifestation of the global flow of labor that Singapore relies on for its economy to work. The female domestic workers from the Philippines form part of the invisible economy. Their labor in the home frees Singaporean women to contribute actively in the workforce, but their work, being limited to the private realm of domesticity, is not always apparent in the visible economy. Orchard Road, being the favorite hang-out of these workers on their day off, also brings them together in the same space as the type of worker occupying the other extreme end of the global labor spectrum—the white expatriate executive, who often has a home in the nearby swanky Tanglin district or just off Orchard Road itself. Adding to the mix of transnational subjects is the ubiquitous tourist on Orchard Road—who is now not identified by where he or she hails from, as tourists from almost every continent can now be found here. They are identified by the slightly offbeat clothes they sport, the multitude of labeled shopping bags and the somewhat worn-out-from-shopping look on their faces. Orchard Road is an embodiment of what Appadurai (1996) terms "ethnoscape", a fluid and shifting landscape of tourists, immigrants, exiles and other moving groups and persons. Here on Orchard Road, the space that almost every Singaporean is conscious of (but may not necessarily frequent), the identity of 'the Other', and thus also of the Singaporean, is constantly re-made and re-defined. A closer look at the Orchard ethnoscape would reveal that these different groups tend to gravitate to different spaces on the street.

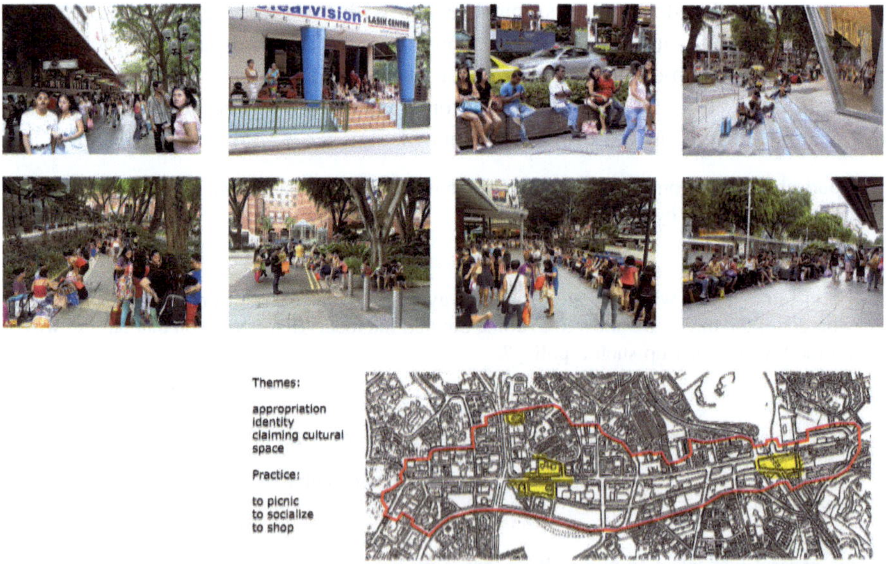

Fig. 3.4 Sunday enclaves

As mentioned earlier, the Filipina workers' core gathering space remains at the promenade outside Lucky Plaza, but emanating from this core, new peripheral spaces in close proximity to the core space, such as perches along the pedestrian path next to Wisma Atria, have sprung up. These peripheral spaces, however, lack the original prominent location and visibility of the core space—instead, they are spaces hidden from view of the main street, and spaces between buildings, spaces which slip through the cracks of programmed spaces.

The Sunday enclaves are bounded in space, in time, and by social distance. They occupy spaces that Singaporeans hardly use—due to a different culture of space use. These groups are somewhat closed in on themselves—the Filipinas keep a distance from Singaporeans, and in fact, seem uncomfortable if the locals approach them. It is when these workers are grouped together that they are empowered with a sense of identity and the right to claim spaces for themselves in the city—in contrast, individual female domestic workers tend to hang out in the margins of spaces, in hidden spaces off the busy sidewalk. The visibility of the Sunday enclaves has generated debates in the press regarding issues affecting these workers.[21] On another level, even with the apparent lack of interaction between the Filipinas and locals in these public spaces, Orchard Road is where such 'silent' groups have a 'voice' in public. The visible culture of space use of the Filipinas gives Singaporeans some degree of exposure to cultures outside of their own.

3.2.4.2 Tribes

Spatial practices, here defined as things that people do which are specific to certain spaces (and times), have always been dynamic on Orchard Road—old networks of bodies, times and spaces have been replaced by new networks which are dependent on new social structures and social practices. I return to the phenomenon of the "Centrepoint Kids" as such a case. During the mid-eighties, the spaces around the Centrepoint Shopping Center had been a favorite hang-out after school for many teenagers. With its central location and prominence on Orchard Road, the spot suited the socializing needs of this group of youngsters, but their presence was then viewed as symptomatic of the social ills of the adolescent demographic. The social pressure from bad press and from youth groups out to 'help' them caused the group to become dissipated, albeit with much resentment, as reflected in this letter titled "Ah Bengs sniff glue, we do not" by 'Former Centrepoint Kid' sent to the Sunday Times Forum in 1986:

> After reading about breaking up the Centrepoint Kids, I've decided to tell you people what Centrepoint Kids are actually. By the way, what you people say about them is entirely wrong! They do not stoop to glue-sniffing or shop-lifting. Those who do these things are school-leavers or bad boys we call "Ah Bengs". A true Centrepoint Kid would go to

[21]For example, that the workers were displayed in employment agencies shop windows were against the practices of a civil society.

> Centrepoint with friends to make more friends. They come from good homes (not broken ones) and are still schooling. Some even come from Singapore's most renowned schools. Normally, they would start by corresponding with each other through the pen-pal column in the Singapore Post. They would also send greeting messages to their friends. Now, do bad boys do that? Centrepoint Kids are well-educated not sluggards. They enjoy this kind of activities because of the monotonous schoolwork, and they need a break.
>
> They usually organise functions every Saturday night. I believe such enjoyment is not against the law. Maybe this news will be very soothing to our ears: There are no more Centrepoint Kids. The majority of them have stopped going there because they understand they will receive lots of trouble by staying there.

Almost a decade after the disappearance of the Centrepoint Kids, beginning from the late 1990s into 2000s, youth culture had found its own niche: what The Straits Times in an 2003 article, called the "Triangle Tribes"—adolescent youths who converged at the "Youth Triangle" defined by The Heeren,[22] Cineleisure Orchard, Youth Park, and Skate Park. Unlike the groups of youngsters hanging out at Orchard Road in the past, the new Tribes that hung out there were in fact embraced by retailers and sanctioned by planners. On any given Friday night at Orchard, sub-species from this group, including the Skateboard Pack, the Heeren J-Popsters, the Cineleisure Railing Huggers, among others, would thrive in this niche of public spaces on Orchard. The patrolling police were wont to leave them to their devices, such as showing off their skateboarding prowess at the Skate Park.

> 'Kids just love to hang out here', says Leslie Teng, 16, who goes to Skate Park every Friday from his home near the Commonwealth MRT station. Of the vibe in the Youth Triangle, he says: 'There are movies to watch at Cathay Cineleisure, stores like HMV and 77th Street at The Heeren, and a skate park. You can get everything here' (Straits Times 2003).

Singaporean society's attitudes towards the social needs of youths have changed to favor their visibility than to encourage subversiveness, and with the setting up of the National Youth Center just off Orchard Road, youths are actively being wooed to the public spaces of Orchard Road. The technology that supports the social habits of the Singaporean youths of today, such as the mobile phone and SMS or text-messaging systems, open up so many choices of where to meet and what to do, such that public spaces must have a critical mass of attractiveness to be the place of choice for the young to hang out. These locations are thus now less subject to chance and propinquity. Labels of where "in" is and which places are "out" are constantly changing for the young and hip crowd that fill up Orchard Road on a Friday night. They are testimonies to the transient nature of these perceptions and the difficulty in pre-programming spaces for specific activities and behavior (Fig. 3.5).

---

[22]The Heeren was popular with youths in the late 1990s into the 2000s, with its retail mix of Japanese-styled street fashion stores and record store chain HMV. The shopping center gradually changed its retail profile, which led to the dissipation of its youthful clientele. This transformation of the Heeren was made complete in 2009, when HMV moved to 313@Somerset, signaling a shift in the geography of the Orchard Road youthscape.

Fig. 3.5 Friday nights

3.2.4.3 Shopping

Although Orchard Road lacks shop windows, the act of buying or having bought something is very evident—people are carrying shopping bags, resting from shopping or buying something from the numerous vendors, pushcart operators or food kiosks along the street. Shopping as an activity is supported by eating. The proliferation of food courts lined up within basements of malls is a typology by itself, as explained by Tang Wee Sung, the then-chairman of the Great Singapore Sale organizing committee in 1997:

> Shopping is not just about price alone. If it were, we'd all be shopping in discount stores and supermarkets. The experience comes in terms of ambience, the way merchandise is grouped together so that it tells a story. Why do the Indonesians still come to Singapore to shop? Because we've got the biggest range of goods around. Where can one get an Orchard Road with a whole string of shopping facilities together on one road, with cafes to stop by in between? I think the heart of shopping is still here. I maintain that is the advantage we have (Dhaliwal and Hameed 1997).

Shopping is more often than not a social activity—shopping with friends or in family groups is the modus operandi. The constituency of the shopping crowd changes diurnally, weekly and also seasonally. The welcome shade provided by angsana trees, arcaded walkways, and covered sidewalks make a big difference to the thermal comfort of shoppers. In fact, hanging around near the entrances of malls to enjoy the leaking cool air seems to be a spatial practice by itself.

In the hot and humid weather of Singapore, places which are either semi-outdoor, such as verandah arcades, or air-conditioned, such as the atriums of shopping centers, are welcomed relief. These places are wells of thermal comfort on the busy shopping street, and as spaces of collectivity, whether owned privately or publicly, they perform roles of public spaces or are regarded as social spaces. Arcaded, in-between spaces are often spaces for resting, waiting and meeting friends. These spaces also vary in the degree of programming—from highly programmed atrium spaces to incidental spaces which act as a buffer zone between public and private. Some degree of improvisation of the use of these spaces occurs (e.g. a vendor may set up a stall in these spaces, or a busker). More often than not, exhibitions or performances, including seasonal events like caroling at Christmas, are held in atrium spaces, drawing the walk-in crowd from the street to the interiors of the buildings (Fig. 3.6). These 'big-city rooms' form part of the network of 'plug-in public spaces' and have a life of their own:

> Fans thinking of pulling a fainting trick on 5566 (a "boy band") should think twice. The boy band is street smart. Tony Sun, the leader of 5566, doesn't think that their fans would pretend to faint to get their attention…His band had whipped up mayhem two Sundays ago at the Atrium@Orchard while promoting their show, Westside Story. The 9,000-strong crowd even jammed up the exits from the Dhoby Ghaut MRT station at one point, leaving angry commuters stuck inside the station. There was apparently a warning issued from the police for the crowd to clear the exits, or the event would have to be discontinued. On the surging crowds, Tony said: 'We can only tell the fans to watch their safety, and threaten to cancel the event if they get too rowdy (The New Paper 1997).

3.2.4.4 People Watching

Coffee culture, exemplified by the outdoor al fresco cafe which has become a ubiquitous feature of Orchard Road from the 1990s onwards, supplies the yuppie crowd with its caffeine shots—the underlying pulse of the new city of bright lights and late nights. The availability of places to sit and watch the goings-on on Orchard Road is a vital ingredient in keeping up the vibrancy of the street. Spaces which require buying a drink to use, such as sidewalk cafés, fast food joints and beer gardens, and spaces which are free, such as public benches, railings, sides of fountains, are used as vantage points for people watching. Different "perches" attract different crowds. With the advent of a coffee culture on Orchard Road, and the proliferation of street-front cafes and coffee kiosks, new vantage points or perches have sprung up on the street, in new configurations and varying degrees of attraction as public spaces (Fig. 3.6):

> Starbucks, Borders, Planet Hollywood and Dome – brand names that roll off the tongue of any Singaporean, young or old, who is in the know. Before you can say "latte", crowds of yuppies and teenagers were flocking to see, and be seen, under those umbrellas as they sipped their macchiatos and mochaccinos at the newly opened Starbucks. Since these new names popped up last year, the stretch along Wheelock Place and Liat Towers is no longer the fringe of activities on Orchard Road. It has since taken a vibrancy of its very own.
>
> Said Pearlyn Yap, 28-year old marketing executive, "Its like a totally new and different culture that has descended on Singaporeans, something which they might have only experienced on holidays in Europe. So it is not surprising that Singaporeans are so attracted to it, because the whole environment here makes them feel as though they are not quite in Singapore, and yet they are" (Straits Times 1997).

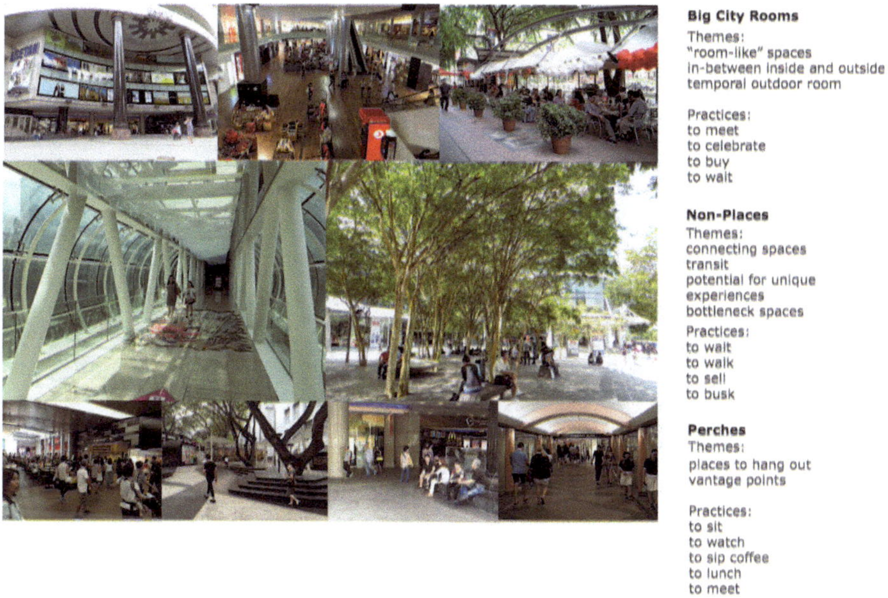

Fig. 3.6 Typologies of space. *Source* Author's collection

Bodies are in passive poses in the café spaces—shoulders hunched, arms around a drink, and the gaze on the street neutral: the privilege of gazing on the street from a safe and comfortable vantage point is coveted and usually paid for. The gazers are often also the object of the gaze—thus bodies are also posed, and well-dressed.

> By far, the most interesting character they've seen is a 'very rebellious-looking guy'. 'He had tribal tattoos all over his body, up to his neck. It's amazing to see people spend so much to put art on themselves', says Ng. Meanwhile, at Borders diagonally across the street, senior manager Christopher Tong says: 'All kinds of hip and happening people come here. Our food and drinks are quite reasonably priced, so that attracts a lot of youngsters. You can see people from all cultures, demographics and colors. It's a cosmopolitan crowd on Orchard Road'.

> Singapore Airlines stewardess Karen Wong, 22, and freelance road-show producer Alvin Lim, 22, hang out here weekly. They spend about two hours and $10 dollars each on each outing. Says Wong: 'It's a good outdoor place, ideal for smokers. It's very interesting to just sit and look at fashion and human behavior. I think people are now less glamorous. They don't take the trouble to dress up any more'. Lim agrees: 'People are more casual these days. No one leaves a deep impression'. Both agree that the wide pavement next to the cafe, which can hold a large number of pedestrians, and the huge video screen across the road make for good lounging.[23]

To the many Singaporeans on Orchard Road, the street is where one might meet someone with a different social or cultural constitution than oneself. On Orchard

[23]From "Tommy", a journalist surveying favorite look-out points on Orchard Road.

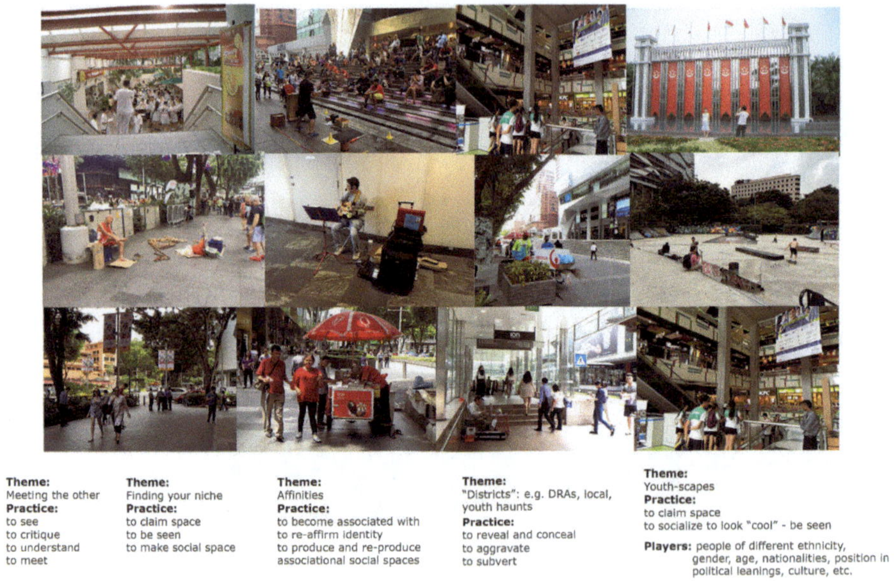

Fig. 3.7 Identities/working

Road, precisely at the promenade outside Centrepoint, is where a young Singaporean child may actually meet someone with hair and eye color different from hers; for the first time, on Orchard Road, a middle-aged housewife might meet a young skateboarder with orange hair and a pierced tongue, while a group of student flag-sellers meet at their favorite rendezvous space (Fig. 3.7).

In fact, because Orchard Road is the kind of public space that enables one to assemble one's own favorite a la carte menu of public spaces, it allows Singaporeans and others of different social backgrounds to meet, more so than any other physical space in Singapore. The dynamic negotiations of space on Orchard Road make this the space of friction,' where ideas of self and of others are constantly evolving. Here, the nature of public space is one in which different spheres exist in proximity at the same time, rather than sharing and using the same space, therefore public space becomes a site of possible exchange and, sometimes, antagonism. Its value as the public space of the nation lies not so much in the sharing of space, let alone a true 'meeting' of different spheres, but in the opportunities offered by urban proximity. An experience of otherness, of spatial practices and views other than one's own, allows for possible 'shifts of perspectives'—something which is not, as is often celebrated in discourses on public space, a pleasant experience, and which at times can even be disturbing or offensive (Hajer 2001, pp. 89–90).

Ultimately, Orchard Road is where the Singaporean "Heartlanders" go out to the city, as described by well-known Singaporean columnist Sylvia Toh Paik Choo[24]:

> ...like you, when I step out on weekends, I don't want to go downstairs of my block or round the corner from my house; why stay in my backyard, I want to go OUT.
>
> And, say what you will, out IS Orchard Road.
>
> All the mallians I talked to yesterday were from out of town ('weekdays heartland, weekends Orchard') and they were out in force (meeting friends) and vengeance (shopping with). All their reasons for joining the thick and the fray boiled down to 'fun lah!' 'shopping!' 'more things to see and buy'. It's not like they can't buy the identical mobile phone accessory from their neighborhood. But there is no frenzy, no carnival, no 'going out' attached to chilling and eating and shopping in your own kampung. Not one young person was bothered, much less aware, that it was crowded, that you sometimes had to push and shove, that for everything you had to queue (The New Paper 2003).

3.2.5 Compression of Scale and Perception

Orchard Road and its public spaces take on different meanings when they are perceived from different perspectives and at different scales. Take for example, the concept of "non-places". French anthropologist Marc Augé defined new places which lack the characteristics of what we understand as traditional places, and termed these "non-places". The term applies especially to transit spaces, or spaces we tend to move through instead of staying in: "Places and non-places are opposed (or attracted) like the words or notions that enable us to describe them. But the fashionable words...are associated with non-places. Thus we can contrast the realities of transit...with those of residence or dwelling; the interchange... with the crossroads...; the passenger...with the traveler..." (1995, pp. 107–108).

However, if we consider the perspective of these "non-places" from the eyes of those who actually use them, we would realize that these spaces are not as one-dimensional as they seem. Here, I would like to illustrate the case through the urban diaries I had kept of Orchard Road, as well as those blogged by other users:

3.2.5.1 'Connecting' Non-places

(From my urban diary—Saturday afternoon)

> 1230hrs
>
> The ubiquitous newspaper kiosks, set up at various points on the street, especially near the exits of MRT stations, are already doing brisk business. The crowd just outside Marriot is

[24]The terms "Heartlanders" and "Cosmopolitans" were coined by Prime Minister Goh Chok Tong in one of his early National Day Rally Speeches to describe the bifucation of Singapore society into two groups, one more home-oriented, the other more globally-oriented.

already filling up the generous covered promenade, stopping to buy morsels of food sold in kiosks.

Going fast down the escalator to the underpass linking Tangs to the Orchard MRT, the weekend crowd is already building up. The blind electric organ player, a permanent fixture, it seems, on the underpass, is playing a melancholy Chinese love song, punctuated by the clinks of shiny coins in his overturned cap.

At the Orchard MRT station lobby, young people are huddled in small groups as they meet, gossip and laugh while waiting for the rest of the group.

1245hrs

While waiting for the light to change at the Orchard/Paterson junction, expatriates in their weekend athletic gear rub shoulders with tourists in batik shirts (from Indonesia?) and immaculately dressed women just coming off work to meet friends at an Orchard Road bistro. On the other side of the road, junior college students in their uniforms hold their donation cans ready for the next wave of people who cross the street at the light, giggling and chatting in the generous shade of the Angsana trees lining the sidewalk.

Two women shoppers, tired from their shopping foray the Takashimaya, sit on the steps of the crescent shaped promenade lining the center, They chatted and whiled away half an hour, just hidden from the frenzy of Orchard Road on a weekend.

(From The New Paper, "Gone in a flash: The mob that wasn't")

At 2.10pm, the Singapore flash mob rode up and down the escalators near the MRT entrance three times. While going down, they shouted 'whee' with their hands raised, as if they were on a roller coaster. The whole thing took about three minutes. By 2.13pm they had split up in three different directions.

'What was that? Some uni (university) orientation is it?' asked 25-year-old Selena Wong who was waiting for her friend at the station. She didn't know what flash mobs were. Flash mobs do not seem to have any political or ideological agenda. They are supposed to be spontaneous, whimsical fun (Narayanan 2003).

(From the Internet blog of "Craig", an American expatriate living and working in Singapore)

I took the train to the Orchard Station, got out, and walked around a spot that Lily and I had agreed to meet. I heard the roar of the crowd as everybody was off from work and going about and Orchard is especially one of the busiest train stations. But then I heard some footsteps running behind me and towards me. I quickly turned around and saw a short (5'4") Chinese girl smiling at me. Lily. "Damn! Craig — I was playing your hide-and-seek game, trying to beat you!"

"You can't beat me babe, I invented the game", I replied. We then embraced and kissed, and then hand-in-hand, walked off into the crowd. We walked through the train station which is of course underground, and then through connecting tunnels and hallways that lead through all kinds of connected underground shopping malls – no, I'm not kidding.

So Lily and I walked through this subterranean shoppers paradise and then popped out onto the street eventually — far away from the original train station. We walked into a small 7/11 store (yep, they even have those here), where Lily bought us both a drink — ice cold cans of "Chrysanthemum Tea", and she also bought me a pack of Marlboro Lights ($6.70).

> We then walked along the throngs of people on Orchard Road and came upon a band of Chinese Acrobats performing on the sidewalk. We sat down and watched them for a bit, commenting here and there of what we liked about their show and what not.[25]

From the above anecdotal accounts of the "non-places" of the Orchard MRT Station and its surroundings, it seems that such spaces provide for quite unique, sometimes intense or novel individual experiences. The spaces described here are indeed transitory spaces but are in many ways 'compressed' public spaces which act as a platform where people with all sorts of different itineraries cross paths and have a chance to spend 'public time'. Staple "fixtures" like the blind man who, with an accordion in the underpass between Tang Plaza and the Orchard MRT, belts out melodramatic Chinese songs, mark the experience of the space as an identifiable place. These spaces are often also integrated with other networks of public spaces and have the potential to become viable social spaces—where one gets to meet, and define the Other.

From the mid-2000s, we have witnessed Orchard Road develop into a space of increased connectivity. However, most of these connecting spaces are invisible from the street level. Today, areas of public space in Orchard Road have been established within a capillary of transitory 'non-place' connections: from interlocking, Lego-like overhead connections between malls, to the extensive network of underpasses connecting the underground train service to the mega-malls of Orchard Road. The multi-street level approach clearly delineates the priorities of access to these individual layers: the underground and above-head connections reserved for human traffic, and the street-level prioritized for the automobile.

The non-street connections in Orchard Road provide shelter, uninterrupted movement and ease of access, particularly for visitors who travel by mass rapid transit (MRT). An underground walkway connects Orchard MRT to Wisma Atria and Ngee Ann City (from its east exit), to Tangs, Marriott, and Lucky Plaza (from its north exit), and to Shaw House and Wheelock Place (from its west exit). Furthermore, Orchard MRT is located within the basement of another shopping landmark, ION Orchard. Most of these connections are open 24-h, such as the walkway located in 313@Somerset shopping center, connecting Orchard Road and Somerset Road.

The incentive to create 'non-spaces' in Orchard Road is palpable; there is now more area per square foot to commercialize. These interconnected spaces, whether suspended in mid-air or tucked underground, are prime locations to capitalize on the crowds channeling through. It is no surprise that the connections in Orchard Road are chock-full of advertisements and pop-up stores seeking to steal the attention of the fast-moving commuters. One mall in particular, Orchard Gateway, has found a way to innovatively benefit from the necessity of 'connecting spaces'. It is a shopping mall that distinctly "straddles both sides of Orchard Road", and is linked by both an underpass and an overhead glass

---

[25]From the Internet blog of "Craig", an American expatriate living and working in Singapore http://www.angelfire.com/ga2/stepstoinsanity/Journal/page6.htm.

Fig. 3.8 Multi-level connections of Orchard Gateway. *Left* Underground MRT Linkway; *Center* Orchard Gateway's Overhead Bridge; *Right* Street level. *Source* Author's collection

tubular bridge.[26] Connections such as these serve as the main arteries to channeling human traffic across Orchard Road, and are prime examples of how non-places in Orchard Road have not only been transformed into 'places', but also 'destinations' (Fig. 3.8).

3.2.5.2 Stories

On an individual level, on the level of small, personal stories, Orchard Road, instead of the glitzy shopping street, can be perceived as a giant playground, a space for individual expression, or even of transgression of the status quo. Sometimes, events that are portrayed on the scale of the individual ring differently from how they are perceived as an urban phenomenon:

(From the Internet blog of "Dexter", 17, a skate fanatic)

> A short word on this area. This is the shopping district of Singapore. There are stuff to skate but human traffic is impossible especially after 5 pm. Lotsa long rails are a temptation to great to resist! Go before 4 pm. Watch out for cops. Prime shopping area means more reason to diss out "injuring passers-by" shit.
>
> Orchard road is one of the best skate spots in Singapore—marble and tile as far as the eye can see. It's got everything, rails, benches, curves, ledges, and about any kind of screwy fixture you can think of. Even though it's awesome skating, you have to watch for the police because you'll get busted faster than holding up a doughnut shop.

(Adapted from The Straits Times, "Jingle bell rock: Orchard Road again but many miss the Swing")

> "Can you take a picture for me?" asked Mr Eddie Teo, 30, an engineer. He was the least unfazed about the lackluster atmosphere of the monthly street closure of Orchard Road for people to use the street to do "their own thing". Others were more vocal with their complaints, "We came here hoping to relive the fun we experienced at Swing Singapore", said disappointed Mohan Chanran, a salesman. "The whole place was so alive then. But tonight, here is the party spirit, where is the fun?"

---

[26]This phrase was used by Orchard Gateway themselves, on their website, to describe the significance of Orchard Gateway's design and location.

> Marcus plonked down his 5 kg mini-compo, sighing, "It's so boring". His group of eight friends were listless. "Why don't you start dancing?" Simon raised his eyebrow. "No thanks, we'd look like idiots if no one else is doing this". Marcus was resigned to the fact that he was going to spend the evening reading a comic book. Maybe Singaporean youngsters are not very good at organizing street parties for themselves… (Lim and Kaur 1988).

(Comments posted on the Internet by Ben Chin, an NUS undergraduate, June 2003)

> Personally, Orchard Road has a special place in my heart as it brings back many fond memories of my secondary school and junior college days when I used to go down after school with my classmates not for any particular purpose but to soak in the unique atmosphere that could only be found there. We would often just find a little corner to sit, where we could bond and play the game "Truth or Dare". The dare would very frequently have something to do with the crowds; for example running around them and shouting some embarrassing phrases or going for a 'dip' in a fountain that is surrounded by many people. Of course, we got our fair share of unwanted attention from the security guards but that did not deter us from having our own fun. At other times, we would simply sit down somewhere to observe the passing crowd, watching the many different kinds of people who have gathered at this place. It is an experience that is not easily replicated elsewhere in Singapore and I am sure that Orchard Road has also left its mark on many other Singaporeans, just like the way these memories have been etched in my mind.

Not all the personal feelings about Orchard Road are positive though—for some, the street only brings on the fear of crowds, and perhaps a tinge of aloofness and loathing of the masses of 'ordinary' Singaporeans there—as if one were a *flâneur*: (From "PJ", a Singaporean, with an overseas graduate degree)

> I hate going to Orchard Road on a weekend. The place is just one big seething mass of humanity. It seems like half of Singapore is walking around there… Half of the people about seemed to be kids with nothing to do. I don't really understand these kids. When I was their age, I had things to do, places to go, people to see. I had hobbies, sports, and when I went out I was going somewhere and moved with conviction and direction. What kind of youth is spent wandering around shopping malls with equally lame friends, window-shopping and conversing with disembodied friends by tapping away at inhuman cellphone keys? Furthermore, they are quite happily willing to pay the obscene prices of shops and restaurants located in a high-rent district. I mean, five bucks for a cup of coffee?
>
> So I shouldered my way past school kids in their uniforms, teenyboppers looking like they were surgically attached (hip to hip) with their significant others, poseurs trying to dress like whatever the fashion du jour happened to be, and other fine examples of Singaporean youth… I stepped past families, skipped over little kids running about underfoot, and pirouetted to avoid a kid who rushed out of a side passage without warning.
>
> Finally I made it to the library, an oasis in the intellectual desert, swopped my stack of books for a new stack, and retraced my steps, past the overpriced boutiques, through the heavily air-conditioned tunnels, reeking with the stench of hundreds of bodies in a way only the tropics can create, and thankfully back to the safety of my car. My throat had long gone dry, my eyes hurt, I felt sweat beading on my back before almost instantaneously evaporating in the artificially cooled air, a heaviness was afflicting my muscles and I felt completely out of place in this land where consumerism, superficiality and artificiality held the upper hand. Twenty minutes later, I was home, and thankfully sank down in my chair with a nice hot cup of tea.

3.2.5.3 Working

While Orchard Road has generally been regarded as a consumption-scape par excellence, there are many who regard the street as 'workspace' (Fig. 3.7). These may be people with regular jobs there or those who would fall under the 'informal economy' in globalization-speak. Ice cream carts—identifiable by their blue umbrellas, and run mostly by old men—seem to be a ubiquitous feature of the Orchard sidewalk, and are in themselves little magnets along the street. In contrast, newspapers kiosks are invariably run by little old ladies in sun hats. Street cleaners are sometimes spotted, but are more often than not the invisible army who keeps the streets spanking clean.

Food vendors, mobile food-truck operators and push-cart operators service the users of the sidewalk—many of these vendors have make-shift tables or stands, and one little old lady actually sits on the floor. Caricature artists display their work on the sidewalk or are seen showing off their skills. Some of these vendors are almost semi-permanent fixtures in a space, that their absences are felt immediately if they are not there. They also serve as points of non-committal contact in the public, and as reference points or even familiar faces in the public, they temporarily define the spaces around them, such as street corners and particular spots along the streets. Their presence is often also dependent on the time of the day, or they may persist from day to night but serve a different crowd—they may serve as time markers or as almost semi-permanent reference points. Here is an account about an old street vendor:

(From The New Paper, "Kopi Auty of Orchard Underpass won't accept handouts")

> Every afternoon the 72-year-old woman takes a slow, painful hour to trudge the 3km distance from her one-room flat on Indus Road to the Orchard underpass beneath Scotts Road. She doesn't take the bus - she can't afford to. Madam Tay earns her money by selling sachets of Super coffee mix and Super chrysanthemum tea, clean towels and aunty blouses.
>
> Walk past her along the tunnel, and you will notice that unlike other buskers or beggars, she never calls nor signals for help. What she does, is smile and hope that someone kind will buy her wares for a dollar or two (Tan 2003).

3.2.5.4 Scales of Engagement

These anecdotes suggest that to perceive the scale of Orchard Road solely from the scale of urban events and phenomena is to overlook personal meaning, positive or negative, that is attached to Orchard Road as a public space. Here, in my attempt to animate personal constructions of Orchard Road as a public space and not only as event space, I have collected, documented, collaged or re-constituted multiple personal perspectives of the urban. The reconciliation of the different scales of the urban intensifies and enriches the resolution of Orchard Road as a public space. It is at this scale of resolution that notions of identity and place can be fleshed out and brought to life.

The scale and urban prominence of Orchard Road allows for the staging of large-scale events and spectacles, often with themes like "health", "nation" or "culture". These spectacles range from nationally televised mass events, product launches, fashion parades by retailers, to the annual Chingay Parade (marking the Chinese New Year) which for a decade-and-a-half were held on Orchard Road, the one-off Millennium Swing Singapore street party, and the short-lived (two-and-a-half years) end-of-the-month road closures for do-it-yourself street parties. The general lack of spontaneity for mass events organized by the state is a well-known fact—these are often played out with coordinated, mimicked movements and rehearsed steps—"following the leader"—with participants dressing and acting the same way. In a performance event, there is often also a strict separation of performers and audience, with a defined realm of activity, bounded in space and time—in other words, staged actions without innate content or meaning that reinforced the status quo (for example, of collectivity as a nation), watched by a sanctioned crowd (Fig. 3.9).

The banal unity and boundedness of these events show a degree of "dis-embedding" of cultures (such as dancing in the street, a concept not indigenous to Singaporean culture) from elsewhere without any attempt at "re-embedding" these within Singapore public space (Giddens 1990). For example, the street-parties organized by the state, which tended to be so bounded with mechanical precision by the allowed time slot such that they seemed to verge on the bizarre:

Fig. 3.9 Mass activities

(Adapted from The Straits Times, "*Shiok?* It's what you make of it")

> At 9.50 pm, men and women were already sweeping the streets, weaving in and out through the dwindling crowd. And seven minutes later, police moved into groups of party revelers to tell them to end their party. The drumming stopped, the dancers stopped, the music stopped and the onlookers gradually dispersed all over Orchard Road. At 9.59 pm, with military precision, the late, late "show" began. It had hints of the closing of a parade, except this time the centerpiece escorted by a traffic policeman on his gleaming motorcycle was a mechanical sweeper. It made its grand entrance as if to show the crowd it was king of the road. Meanwhile, at the Dynasty Hotel end of Orchard Road, five other traffic policemen sitting neatly in a row on their revved-up bikes waited for their turn to wrap up the party.
>
> It came exactly at 10.02 pm. With the five motorcycles coming down the road, the crowd started to disperse – some into the MRT stations, some milling on the pedestrian mall. By 10.12 pm it was all over, and the motorists were back to showing the party-goers that Orchard Road belonged to them (Tang, Tan and Pow Chong 1989).

However, once the road closures were strictly for do-it-yourself street parties, the concept shriveled up and died because Singaporeans do not have such a spatial practice ingrained in their social culture.

3.2.6 *Narratives of Loss and "Absence"*

Orchard Road seems to attract its own "discourses of nostalgia", although there was really a dearth of built history (Berman 1997, pp. 76–83). Most of the voices of lament seem to decry the loss of an age of innocence, the loss of nature or of authenticity.

3.2.6.1 Angsana

The most vociferous of laments come from the group which we will call the "retrogrades" (Sagalyn 2001, pp. 457–462). The retrogrades are often those who equate the loss of old places and spaces with loss of personal memories, and are likely to belong to older generations of Singaporeans, and act on these sentiments by being absent from Orchard Road—it is a fact that the local crowds thronging the street are mostly of the young and the hip. They lament the loss of an authentic past, are suspicious and skeptical of planned changes, and ruefully view some of the new schemes as careless or irresponsible in its treatment of history. Some would prefer that places like Orchard Road be allowed to develop 'naturally' and perhaps die a natural death, rather than suffer from any form of intervention. Here is an example of such a sentiment:

> In earlier times, the area around it would have been more green and even more striated. Orchard road derives its name from the nutmeg and gambier plantations owned by colonial planters…in the middle of the 19th century. These plantations were at the edge of virgin jungle and a tiger was even spotted in 1846. Such was the mystic of Orchard. But the

> plantations and the jungle are gone. The only 'tigers' along Orchard Rod these days are the ones you down alfresco at its many outdoor cafes and bars. The angsana trees are the ersatz jungle that are enjoined by more systematic green and exotic "landscaping" co- opted to evoke another mystique – the garden city or the city as garden. But these too may go (Tan 1999, pp. 94–99).

The author of the above comment is a practicing architect in Singapore who decries the loss of a greener, more 'natural' state of Orchard Road as exemplified by the shady angsana trees, as opposed to the highly abstract and constructed environment of today. He goes on to say:

> At this very moment another narrative is being constructed. The future of Orchard Road is being plotted by the Singapore Tourism Board (STB). Soon, it is rumored, the angsana trees will be gone; there is an ominous campaign of extermination of the angsana in the face of this island, because they are difficult to 'maintain'. Instead, a re-forestation of the past may take place. And soon the orchards may perversely reappear again, if only as a theme. Apparently still not vital enough for them, the STB together with foreign consultants, local architects and planners are huddled together to conceive an immaculate rebirth of Orchard Road (ibid.).

3.2.6.2 Tabula Rasa

An even harsher critique comes from influential architect and theorist, Rem Koolhaas, whose comments refer to Singapore in general, but I apply them here to Orchard Road in particular, as he refers to "the global consumer frenzy" that had "perverted Singapore's image to one of caricature: an entire city perceived as a shopping center, an orgy of Eurasian vulgarity, a city stripped of the last vestiges of authenticity and dignity" (Koolhaas 1995). The reason that Singapore is "a city without qualities", according to Koolhaas, is because of the "curse of the tabula rasa", such that "once applied, it proves not only previous occupancy expendable, but also each future occupancy provisional too, ultimately temporary. That makes the claim to finality—the illusion on which even the most mediocre architecture is based—impossible. It makes Architecture impossible" (ibid.). Koolhaas' charge is that the ease which Singapore's built history has been erased makes it impossible to have a legitimate built history and thus authenticity as a city. Even worse, while Tan ruminates that the colonial orchards may yet reappear, albeit as a theme on Orchard Road as a kind of "re-forestation of the past", Koolhaas asserts that "history, especially colonial history, is rehabilitated, paradoxically because it is the only one recognizable as history".

With regard to lost nature, one might perhaps pick up the guidebook compiled by two serious naturalists in Singapore, one a professor and the other a nature guide, on the "Nature Guide to Orchard Road" detailing the flora and fauna found on Orchard Road (Nathan 1994), itself a nod to the resilience of nature, which is not limited to angsana trees. Although it is true that the colonial history of Orchard Road had given the street the legacy of its name, we should well remember too, that the monoculture of nutmegs cultivated by the colonials was what led to the disaster that caused the sudden death of the plantation era. The lesson in history being that diversity of (crop) culture would have ensured the continued vigor of the venture. This lesson is still applicable today to the new plans for Orchard Road: that its very survival would hinge on the continued diversity of activities, architecture, culture, people, goods, public spaces and of memories.

Nostalgia takes different forms—there are those whose nostalgia take a more positive turn, so that while they are wistful and sentimental about memories of a more innocent age of Orchard Road, they realize that those memories belonged to a different time and that their loss is inevitable—they may even view change as progress. These include the "reminders" and the "resilients" who may recall fondly the Mont D'or and Tivoli café which were famous for cream cakes, shopping at the old C.K. Tang or going up the first escalator in Singapore to watch movies in the old Orchard Cinema. For some of these people, who may now be middle-aged or senior citizens living in the new towns, the reasons for going to Orchard Road dim with the passing of each landmark or public space which would have changed.

However, they are optimistic in a pragmatic way about the future of Orchard Road. Their nostalgia takes the form of memory trips to the past, associating people and places—but they do not miss the lack of good public transport or the persistent floods that plagued the area. Their view comes with the knowledge that their loss is personal and if it exists, the "golden days" of Orchard Road is a shifting ghost of a memory. For them, it is clear that nostalgia is a kind of luxury that can be enjoyed only because there are new frames with which to regard the past fondly and at a distance, without the accompanying dirt and grime.

Nostalgia for a remembered or idealized past is not part of the construction of Orchard Road's history, but rather a state that is itself symptomatic of the post-traditional condition of the present time and place in Singapore. With Singapore's rapid growth manifested in its built environment, the decade of the 1980s to 90s was a "historical moment" of the arrival of the "nation" in the economic sense. But the city became somewhat removed from its fleeting past, which some had perceived as reflective of a certain intangible lack of spirit and soul (Kong and Yeoh 2003, p. 132).

3.2.6.3 'Absence'

For all of the presence of the diverse entities who converge on Orchard Road, the 'absence' from Orchard may sometimes go unnoticed. Certain slices of the Singaporean demographic continue to be under-represented here—e.g., one would hardly find the Malay housewife, the retired taxi- driver, or for the matter, the 'coffee-shop-auntie' from Ang Mo Kio New Town, on Orchard Road. The attraction of Orchard Road fails to lure these types of people. In fact, it may actually repel some of these people. Why is this the case?

The types of programs offered by Orchard Road as well as the constant re-invention of itself tend to alienate those who are less likely to splurge on shopping, or to pay sky-high prices for a cup of coffee—i.e. to partake of its many 'paid-for' pleasures, and thus have limited 'access'. The types of events organized for Orchard Road tend also to attract the young or the fit, such as large-scale fashion catwalk shows, street parties, mass exercises, product promotion catering to the youth market, and so on. These slices of the demographic are alienated by the fact that there are hardly any reasons for them to be there. The street that they remembered from times past is no longer the same street of the Dolce&Gabbana crowd or of Swing Singapore. The public 'realms' which have established themselves on Orchard Road are increasingly defined in such a way that the "heartlanders" may no longer find their place.

One of the reasons that "heartlanders" still venture out to "Orchard" is the open-house at the Istana on national holidays, where long queues offer sights of Muslim women in tudungs with their brood in tow, the bespectacled Indian grandmother in her traditional garb, as well as other people from the new towns who are less often seen there.

However, the projected development of Orchard seemed to offer more of the same. The Singapore Tourist Board and the Urban Redevelopment Authorities, who have been charged with implementing the "Rejuvenating Orchard Road Plan, 2005" have continued to fiddle with Orchard Road, in the typically Singaporean bid to rival other great shopping streets in the world.

Announced Minister Vivian Balakrishnan[27]:

> Orchard Road can be likened to a stage, where retail, food and entertainment businesses, events and people interact and thus deliver a compelling performance… Orchard Road must constantly offer new experiences by weaving together the latest in design, fashion, entertainment and cuisine into memorable storyboards.
>
> Events in particular will be a critical element in the rejuvenation of Orchard Road. Events add vibrancy and color and make the city a more interesting place to live and visit. Every year, Orchard Road plays host to several key events, such as the Fashion Festival, Great Singapore Sale and Chingay. We must do even more to develop an exciting array of events and festivals that will go on all year round.

3.2.7 Constructed Authenticities and Their Legitimacy

If we put aside discourses of nostalgia mourning a lost history, what then are the new mimetic structures of Orchard Road? What legitimizes its history if it exists in a city described by Koolhaas as "perpetually morphed to the next state"? In other

[27]From Dr. Vivian Balakrishnan's speech at the press conference on 'Sharing the Vision for Orchard Road's evolution', 29th March 2005. In the speech, he also noted that 'the Prime Minister had earlier announced that STB will spend $40 million to revamp Orchard Road infrastructure' and that the 'MOF has approved a suite of incentives and the URA has relaxed development rules and guidelines'.

words, how do we construct a new authenticity for Orchard Road? To answer these questions, we have to return to our discourse on Orchard Road and the compressions of time and images, of people and places, and of scales of perceptions.

3.2.7.1 Constructions

The sedimented history of Orchard Road avails itself in a closer look: its past lay buried in the names of streets and buildings,[28] in its collaged pavements of different vintages, in the course of the Stamford Canal which runs quietly and almost invisibly, and in the old angsana trees gracefully flanking the sidewalks. Refurbished exteriors of old buildings may give a new identity to Orchard Road, but their memories are old and there is most certainly a layering which even the new masterplan (STB 2001) for the revitalization of Orchard Road would not cover.

The dialectic of people and identities on Orchard Road reflects the global reach of Orchard Road, and such diversity will ensure that it is alive and well. Orchard Road does not cater only to the shopper or consumer—it continues to be a public space, and even a workspace and home for some. The scales of events on Orchard Road, ranging from the adventures of a young skate-boarder, to collective urban experiences like celebrating Christmas at Orchard, reflect the variety of perspectives that form the mimetic structures of Orchard Road—from the personalized to the spectacular.

Narratives of loss simply do not apply to Orchard Road. Instead of a systemic "erasure of history", the mechanism at work here is more akin to the constant construction of the hyperreal—what Baudrillard describes as the creation of a model for which the original does not exist (Baudrillard 1994). Ask anyone who knows Orchard Road what immediate imagery comes to mind, it is more often than not the ersatz concrete and glass pagoda-like tower of the former Dynasty (now Marriott) Hotel and the curving green Chinese roofs of its Tang Plaza podium. This iconic complex is the re-incarnation of the old flagship store C.K. Tang Building (1958), which was demolished in 1982. It is now supersized, with its signature Chinese green roofs magnified and multiplied a few times over. It is no secret that studies have found that Singaporeans, "despite being aware of other highly imageable streetscapes in the world, still think that the Orchard-scape is aesthetically pleasing" (Yeung and Savage 1995).

Notice also the brass embossed plates depicting local fruits as you walk on the pedestrian mall. Whether the fruit trees that lined the nutmeg plantations existed or not is beside the point, as it is the romance of the fabled fruit orchards that counts.

[28]For example, the Wisma Atria building sits on the site of the former Indonesian Embassy. The Ngee Ann City Complex as well as Wisma Atria sit on land owned by Ngee Ann Kongsi, a Chinese clan, which owned the parcel of land previously known as Tai Shan Ting, a Chinese burial ground. The cemetery was cleared in 1957 and leased to the present Meritus Mandarin, Cathay Cineleisure Orchard and the Indonesian government. See also http://www.ngeeann.com.sg/webtop/property.phtml.

The simulated winter scenes which come right out of a Hallmark Christmas card that pop up on Orchard Road every November have definitely not existed here, but are now regarded as part of the seasonal change of 'weather' on Orchard Road. Even the air-conditioned food courts that proliferate in the buildings on Orchard Road are a hyperreal quotation of the rowdy open-air food stalls of the old long gone carpark on Orchard Road.

In the conservation mode of the 1980s, the Straits Chinese-style arcaded shophouses of Peranakan Place and Emerald Hill were rehabilitated as a cultural showcase of Peranakan lifestyle and culture. This venture to sell the 'culture' of the Straits-Chinese who ironically were themselves adept at adjusting culturally to their new environment, failed as the museum and Peranakan themed restaurants were not significant enough draws to be economically viable. Now, these buildings house electronic goods stores, popular watering holes and al fresco cafes within their Old World facades. One thing is for certain—Orchard Road has none of the self-conscious discomfort that comes with the need to preserve itself like a museum. It embodies instead a kind of reckless ambition to move ahead of itself.

3.2.7.2 Flux

In the end, we return to the analogy of Orchard Road as a channel, or conduit: that through which history flows through the present and its future trajectories, sedimenting, eroding. But its banks are fertile because of its active life as a mover, not a stagnater; that which collects, brings together, mixes the different lives which inhabit its waters, sometimes bringing into friction, at other times bringing into happy collisions; that which distributes, shares, transmits, and conducts the energy, the life-pulse and the wealth of the city. It is a life-line for some, an occasional thrill for others; it is that which connects—for Orchard Road is intimately connected with the surrounding city fabric—it remains an important thread that connects the heart of Singapore to the Marina Bay Downtown, and to the waterfront. It still is an important commercial corridor and connective tissue which links recreational, civic and cultural nodes in the city. Plans have been proposed variously in the past to divert vehicular traffic and pedestrianize Orchard Road, but these have never been found feasible, for Orchard Road cannot be reduced to an isolated urban shopping mall—that would have sucked out its life-blood and sounded its ultimate death knell.

The wide sidewalks on Orchard Road had pre-empted the lively shopping street, just as the lively shopping street pre-empted efforts now to "make it more happening",[29] One such effort is Pedestrian Night, a monthly occasion where Orchard Road—specifically between Scotts Paterson Junction to Bideford Junction—is

[29]The Singapore Tourist Board (STB) and URA had worked together to showcase an exhibition, titled, "Make Orchard Road More Happening!" These plans were featured in the March/April 2001 issue of URA's *Skyline* publication.

Fig. 3.10 Pedestrian night. An overview of Pedestrian night, 2nd January 2016. *Source* Author's collection

closed to cars. Since a successful six-month trial period was initiated in 2014, on every first Saturday of the month, from 6 to 11 pm, Orchard Road transforms into a pedestrian-only thoroughfare that boasts street performances, live music, various promotional booths and prize-giveaways. Pedestrian Night is hosted by the Orchard Road Business Association (ORBA) and the Singapore Tourism Board, as part of their plans to revitalize Orchard Road as 'a great street'.[30] According to ORBA, the first six Pedestrian Nights' attracted an average of 50,000 shoppers, twice the average Saturday night footfall (Straits Times 2015). The street closure, for a stretch of about half a kilometer, is organized thematically, with each occasion having a specific theme, such as mass tennis, yoga or 'Movie Night'. However, the monthly street closures faced an uncertain future as the novelty of Pedestrian Night wore off, and retailers felt that the increased footfall did not translate to increased sales (Channel News Asia 2016) (Fig. 3.10).

Orchard Road's public space can be understood using two concepts: 'levels' and 'barriers'. A closer inspection of Orchard's street-level cross section reveals different levels of the ground. The street level is the lowest and belongs to the car, with only a handful of pedestrian crossings on the major intersections of Orchard

[30]'A Great Street' was Orchard Road Business Association's tagline for Orchard Road.

Fig. 3.11 'Levels' and 'Barriers' of Orchard Road. *Source* Author's collection

road. The next 'level' is the pavement, belonging to the pedestrian. The road level and pavement level are buffered by a stretch of barriers: shrubs and metal railings, as signage reminds pedestrians of their limits to the space and diverts them to alternative connections overhead or underground. The final (and highest) level is that of commerce, where shopping centers stand. On Pedestrian Night, the amount of public space that is available to the pedestrian is temporarily extended. However, the same barriers that deter jaywalkers from entering the fast-moving traffic on a normal day also prevent the flow of people from the sidewalks to the street and vice versa. Due to this, events like Pedestrian Night tend to create new event spaces on the street as a contained space of itself. As such, despite the intention behind Pedestrian Night to 'inject street vibrancy' and to portray Orchard Road as an 'integrated community space', it remains a largely segmented and isolated event. However, as the Pedestrian Night initiative suggests, Orchard Road is a space that is constantly being re-thought and re-configured to create a version that seeks to better its former self, albeit to varying degrees of success (Fig. 3.11).

The self-conscious tinkering of the notion of Orchard Road and what it should entail often comes after the fact that it has somehow come of age, though no one could put his finger on what exactly made it tick. Although those who have power to intervene with the development of Orchard Road seem to wield the power to decide its fate,[31] Orchard Road and its public spaces, like the conduit that we imagine it to be, takes on a life of its own. The legitimacy of the post-traditional environment of Orchard Road lies in its ability to flow with the times and constantly re-invent itself. In many ways Orchard Road is a hyperreal environment, in which reference is made simultaneously to its imagined past, to the reenactment of its real history as a commercial and entertainment center, to abstracted models of great shopping streets borrowed from elsewhere in the world, and to some extent, to the notion of the urban shopping mall. Narratives of loss do not apply in its rapidly

[31]Those who wield power to change the shape of Orchard Road include the Urban Redevelopment Authority (URA), the Singapore Tourist Board (STB), the Orchard Road Merchants' Association, the Land Transport Authority (LTA), among other government, quasi-government and private interests groups. In order to help propel Orchard Road in its future development, the authorities are considering implementing the concept of a Business Precinct Management Corporation to provide a vehicle for stakeholders to take ownership for promoting their business precinct and to pursue projects that add value to the precinct as a whole.

changing face, but its continued relevance to the hearts and minds of Singaporeans lies precisely in its abilities to morph into new, constructed authenticities, and in its capacity and potential as the public space of the nation.

3.3 Little India: Spaces of Ethnicity, Exchange and Boundedness

Serangoon Road[32] had been one of the earliest roads to be built in Singapore, first appearing on the 1828 Plan of Singapore[33] as an unnamed road leading between the Town and a harbor in the north. Its construction was closely tied to the geography and land use of the area, which became rapidly cultivated. The area's abundant supply of grass, water and also labor available from the nearby Indian convict prison quickly attracted many south Indian cattle farmers. Although not designated by the British colonials as an Indian enclave, the area had by the 1880s grown as a center of south Indian economic, social and cultural life. A brick kiln industry, livestock traders, butchers, dairymen, manual laborers and retailers had set up shop. Although by the 1930s, the cattle industry had declined, the area continued to flourish as a commercial and residential hub for the Indian community. Shops, houses, temples and mosques made up the urban conurbation, which is today known as Little India.

While the 1970s and 80s saw active slum clearance and the insertion of public housing into this part of the city, the 1980s and 90s were marked by conservation efforts by the Urban Redevelopment Authority (URA). With the lifting of rent control laws by the late 80s and the push for economically viable conservation policies, some degree of gentrification is evident, but the intensively used retail and residential properties remained largely intact. Today, Little India stands as a unique and culturally active area and is regarded by the Indian community in Singapore as the social space of the now dispersed Indian population.[34] Unlike Chinatown in Singapore, which many think of as a thematized, touristic or fossilized district (the Chinese community in Singapore is truly dispersed and Chinese goods, practices and social spaces are spread throughout the island), Little India remains the nub of "Indian-ness" in Singapore. Little India is thus perceived as the public space of the Indian community in Singapore, the space for the production of social, cultural and spatial practices of the ethnic group (Fig. 3.12).

---

[32]According the URA's Serangoon Planning Report 1995, "Serangoon Road was named after 'the Rangong', a bird of the stork species which used to flourish in the swamps around the Serangoon River in the old days".

[33]This plan was attributed to Lieutenant Phillip Jackson.

[34]It is a policy of public housing to achieve an ethnic distribution that does not concentrate any ethnic group within any district, even at the scale of an apartment block.

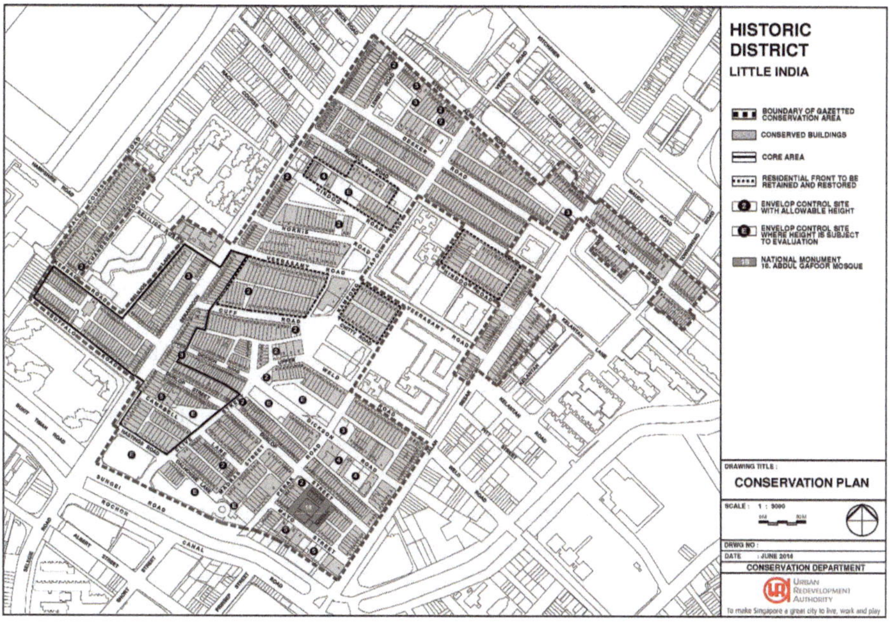

Fig. 3.12 Little India conservation plan. *Source* URA library archives

3.3.1 Framing Little India

For the purposes of this study, the area referred to here as Little India is bounded by the Rochor Canal in the south, Jalan Besar in the east, Kitchener Road to the north and Race Course Road in the west. However, the study is focused on public spaces rather than the fabric of the district in general, so some areas will feature more prominently than others within its general environs.

In terms of physical urban spaces, Little India has few open plazas or deliberately designed public spaces. The fabric of the ethnic quarter is primarily composed of long block forms of 'shophouse' units, with insertions of compact public housing. Serangoon Road runs through as the main artery and small-scale perpendicular side roads lead off the main street. The resulting types of public spaces tend to be more incidental and are 'found' or 'stolen' where available: streets, the five-foot ways, lane alleys, spaces in between buildings and on green undeveloped sites. The conglomeration of these spaces make up the public space of Little India, and it is in this field of physical spaces that I will examine the spatial practices that embody concepts of ethnicity, exchange and boundedness. The disparate nature of public spaces here and their interstitial locations act somewhat like an organic weave of instances that are accommodated as the area grew and densified.

Historically, the district had flourished through opportunistic developments and had grown and adapted around geographical conditions: its central location provided an abundance of water and accessibility. As a working ethnic quarter in Singapore, it serves not only the local Indian community, but also plays host to multitudes of transient workers from south Asia. It is also at the same time, a major tourist destination in Singapore. The role of its public space is thus not limited to that of an ethnic quarter, but it also reflects the impacts of the globalization of labor, and of the interactions of the various entities that inhabit the quarter. In particular, this chapter will focus on three salient aspects of its public space:

- As the space of representation of culture, and the production of identity through spatial practices. The coterminous space of history, geography and the contemporary Indian community are overlaid here in public space.
- As the space of exchange—of economic, social and cultural entities, of goods, services and information. Ideas of "front" and "back" activities in public space are constantly shifting and are framed within notions of gendered space, space for tourist consumption and space of subversion. The constantly shifting definitions of "insiders" and "outsiders" would also be discussed.
- As spaces of boundedness in its many translations: the locale, but a fragmented locality; heavily used threshold and interstitial spaces due to the lack of open spaces; temporally-bounded public spaces, used only as such within time-defined limits; spatially-bounded public spaces, where contestations of space have led to real or imagined boundaries and bounded-spatial practices. The state's presence in public space, the form of control it exerts over behavior in public space, as well as negotiations over space use and correct behavior in public are examined.

3.3.2 Little India—A Brief History

Growth around the area now known as Little India started around the 1830s with *padi* fields and vegetable gardens flanking both sides of Serangoon Road, the street that connected the settlements in the south with Serangoon Harbor in the north.[35] Its proximity to the European Town led to the development of a Race Course in the vicinity in the 1840s. From around the 1850s, the natural predisposition of the land to an abundance of grass, as well as cheap labor from an Indian-convict jail in Bras Basah favored the burgeoning growth of the cattle industry in the area. This attracted many new south Indian immigrants to Serangoon such that as the cattle trade grew and replaced agriculture as the main economic activity of the area. Because of its central location, the area also grew with the establishment of medical institutions such as the Kandang Kerbau Hospital, an asylum and a Government

[35]For the history of Little India, I have consulted several sources, in particular the URA's *Historic District: Little India* (1995) and Sharon Siddique and Nirmala PuruShotam's (1990) *Singapore's Little India—Past, Present and Future* for dates and information.

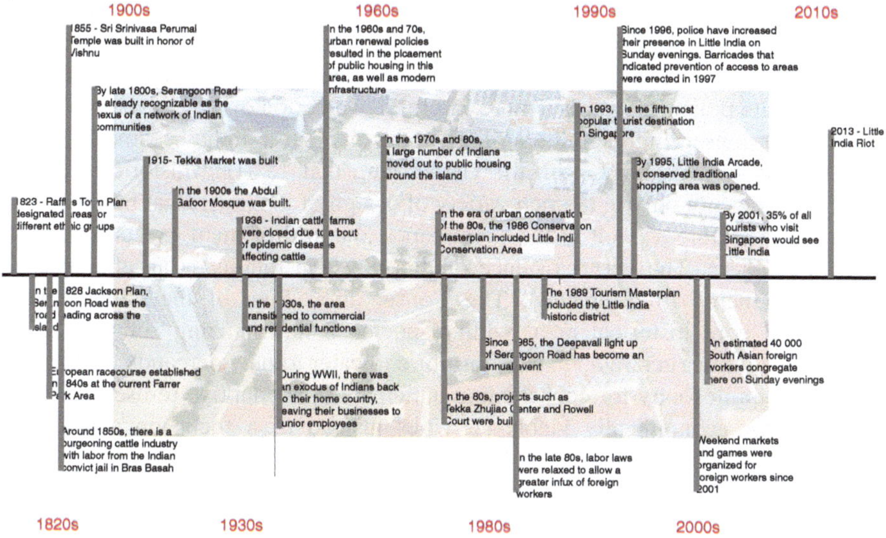

Fig. 3.13 Little India Timeline

Dispensary. By the end of the nineteenth century, the Serangoon area was a hub of the Indian community, with many goods and services set up to meet their needs, and ancillary trades that support the cattle industry (Fig. 3.13).

Although the cattle industry was phased out by the 1930s due to a world-wide cattle disease that affected animal stock, the beginning of the twentieth century saw the continuation of the cattle boom, consolidating the trade under the control of a few enterprising individuals, such as Kader Sultan, also known as the "Cattle King of Serangoon Road" (URA 1995, p. 17). Many new shophouses were built during this period to house the increasing numbers of predominantly Indian businesses and services, which were to eventually replace the cattle industry in the pre-war era. Religious buildings such as temples and mosques sponsored by patrons and businessmen were built from the late 1880s to the early 1900s, and the Tekka Market in 1915. Only the onset of World War II from 1941 to 1944 interrupted the growth of the area. As many business owners and merchants returned to their native India, businesses and shop premises were sold to Indian workers who stayed on in Singapore.

By the 1960s and 70s, the building boom in the area was over, and while businesses had prospered in the post-war era, the upward social mobility of the Indian families led to a high degree of exodus from the area, with these families moving into the public housing of the New Towns. The district became more of a commercial area, with rudimentary dormitories on the upper floors for newly arrived Indian workers. Older parts of the quarter were also torn down during this period, for the insertion of public housing blocks with multi-storey carparks and commercial podium areas such as Zhujiao Centre (now Tekka Center) in 1981,[36] and with these, the influx of non-Indian residents into the area.

By the 1980s, after a decade of rapid development and urban renewal, there was a shift of state policies towards a more careful reevaluation of new development and the need for conserving the built heritage of the city. Little India became a gazetted conservation area in 1989, with the URA citing that "...(t)he historical value of Little India lies not only in the rich variety of design of the individual buildings, but in the urban texture as well as the streetscapes, the grid of main streets, side roads and open spaces. The very existence of this physical totality is an eloquent historical statement" (ibid., p. 23). The Conservation Plan of the ethnic quarter identified a core historical area, with plans for pedestrianizing some streets, a bazaar, new infrastructure and street furniture to cater to new demands for the area. The URA's philosophy for conservation anticipated the adaptive re-use of existing structures for economically viable activities instead of a focus on a wholesale preservation of the district and its businesses.

With the eradication of the Rent Control Act so that owners of properties would be incentivized to refurbish their properties, many businesses and residents were evicted from the area. However, business premises and the sleeping dormitories above shops for workers became intensively utilized such that many of the traditional businesses in Little India had remained intact and continued to thrive. One of the key adaptive re-use projects here had been the Little India Arcade project. Three city blocks of shophouses had been adapted into a festival bazaar, which opened in 1995, forming a notional gateway into the district (Chang 1996, p. 145). While changes such as these led to a turnover of the traditional community of the quarter, businesses and residents outside of the core conservation area remained little affected by the conservation plans.

Little India today is far from the swampy vegetable farms and cattle pens of the nineteenth century, but it is not so far removed from the "complex community network derived from the 1930s resident-commercial base, wherein cottage industries continue to operate quite independently of the highly modernized sectors of the economy" (Siddique and PuruShotam 1990, p. 81). The historical urban fabric of the quarter is largely retained—this had begun as a linear settlement along Serangoon Road, then the cattle pen areas (*kendang kerbau* in Malay) northwest of the road, and then eventually the drained swampy areas southwest of Serangoon Road, which originally comprised small-scale private access roads bearing the names of old colonial property owners. Shophouses[37] in Little India look less ornate and more utilitarian as most of these were built during the early phases of its history. The two main streets that run through the district are Serangoon Road and

[36]Other public housing developments include Rowell Court (1982) and Kerbau Road (late 1980s).

[37]These are terraced 2, 3 or 4-storeyed townhouse-like units built contiguously in blocks, separated from each other by party-walls and bounded on the front by streets and on the back by backlanes or alleyways. Blocks may vary in length from 20 m to more than 100 m, depending on the location. The side adjacent to the street has a cut-in arcaded walkway, called the 'five-foot way', allowing a thorough passage within the block.

Jalan Besar. Most of the smaller-scale streets run perpendicular to these streets, while some are skewed at acute angles, as these were originally roads leading to private estates.

Little India plays an important role as a tourist attraction in Singapore, with nearly 20 % of tourists[38] to the country visiting the ethnic quarter. Every year since 1985, a street light-up festival has been held in October, to coincide with Deepavali or the "Festival of Lights". While the Singapore Tourist Board's (STB's) colorful brochure for Little India presents the area in the form of tourist walking routes through the district, a small note at the end warns that "...on Sunday, Little India can become rather crowded when many of the Indian workers go there to shop, eat, or simply catch up with friends" (STB 2002, p. 21). The statement refers to the phenomenon of an estimated 30,000[39] temporary workers from the Indian sub-continent converging on the streets and public spaces here from the late afternoon of every Sunday.

> ...Little India, with its many Indian shops, restaurants, temples and mosques, became a natural attraction to these workers. It has served as a familiar landmark and meeting place for many foreign workers during their off day... This poses a tremendous problem to the residents in Little India. They face problems of road congestion, littering, noise and obstruction (Lee 1999).

Little India is an enclave for transient workers, specifically migrant laborers from the South Asian subcontinent. This present-day reality has been shaped by immigration policies that began to liberalize in the late 1980s, primarily due to the need for low-skilled labor due to the expansion in the building and construction sector in Singapore. Today, in parallel to the well-educated local population taking up jobs in the professional, management, executive and technical (PMET) sectors, transient worker inflows into Singapore have also grown exponentially, leading to the number of non-resident workers growing from 311,300 in 1990 to 754,500 to 1.63 million in 2015.[40] More than ever, now there are more transient workers across the island who share a need to have access to public space. This had led to new and sometimes volatile spatial negotiations, which, in a single instance, had culminated in a riotous manner.

In December 2013, an anomalous event provoked the need for public space in Little India to be re-examined, re-negotiated, and eventually, re-ordered. A fatal accident involving a bus and a Tamil migrant worker at the junction of Race Course Road and Hampshire Road precipitated an angry retaliation by fellow Tamil migrant workers gathered in the vicinity. Law enforcement personnel were caught off-guard by the chaos that ensued, which resulted in emergency vehicles overturned and set ablaze amidst the turbulent protests. While no residents or civilians were harmed, the incident cast a national spotlight to the large number of transient

---

[38]2013 Singapore Tourist Board (STB) statistics.

[39]This number is an estimate from bus companies that ferry workers from their dormitories to the Serangoon Road. *Source* http://www.straitstimes.com/singapore/little-india-home-away-from-home.

[40]Statistics from *Key Demographic Indicators 1970–2015*, the Department of Statistics, Singapore.

migrant workers who frequent Little India. The incident was also a jolt out of the blue, as street unrest of such a scale had not been seen in Singapore for decades.

Subsequently, state-controlled media outlets labeled the incident as the "Little India Riots", and attributed the unruly behavior of the migrant workers involved to excessive alcohol consumption and a mob mentality. Local residents flooded social media and news outlets with xenophobic comments and recommendations on how public space in Little India should be altered by imposing stricter controls to hinder migrant workers from crowding the area (Straits Times 2013). The police reacted by imposing a temporary ban on alcohol sales, and demarcating a subsequent 'dry zone' within Little India, where the sale and consumption of alcohol was prohibited by strictly enforced law.

Although the Little India Riots was an isolated occurrence in the longer history of public spaces in Singapore, the incident has impacted urban management and the conceptualization of public space in Little India. A fine-grained analysis of spaces and spatial practices in Little India, which includes an account from the perspective of a migrant transient worker, will help us to understand the context of everyday life and its changes in the ethnic district.

The urban fabric of Little India is finely and tightly organized, with public spaces confined to the streets, back lanes, five-foot ways, spaces in front of temples and mosques, open car parks and green parcels of reserved sites or infill sites, as well as the semi-public spaces around Tekka Center and the Mustafa Center, building perimeters and the semi-private spaces of the public housing complexes. While at first sight the public spaces here seem neutral and nondescript, they are far from abstract in nature when we examine them within the contexts of everyday life, and understand these as living, social spaces. The spatial practices that overlay the spaces of Little India offer multiple narratives of public space that present a complex account of the everyday life of the urban area that ultimately provides us with an understanding of the operative nature of public spaces within such a context.

3.3.3 Space of Representation of Culture

Little India on the level of material culture, had been described as the hub of Indian social space:

> Serangoon Road is the locus of Singapore Indian everyday and festive culture. In the first place, the very use of and behavior in Little India follows, predominantly, Indian cultural patterns. Examples include the style of space use in shophouses including the ubiquitous framed favorite deities most often occupying wall space close to the cash registers; patterns of male-female relationships where there is greater conformity to maintaining physical distance … and the tendency to use Tamil/Hindi or other Indian languages for communication. In the second place, one finds in Serangoon Road all the necessary artifacts and implements – ornaments, deities, saris and so forth – without which one cannot fully… participate in being Indian… Most significantly, Little India and the Indian things one can purchase there are used for the most part as necessary aspects of one's culture by most Singapore Indians (Siddique and PuruShotam 1990, pp. 81–82).

Despite its role as the physical repository of Indian culture, the district has few buildings featuring Indian architecture or urban form reflecting south Indian traditions. However, what the district offers to the Indian Diaspora is tangible, material culture, such as Indian artifacts, foodstuff and cuisine, religious buildings and popular culture and media, as described here in a travel article:

> You can buy anything Indian here, just like in any Indian bazaar street - the choicest silk sarees, elaborate Indian gold jewellery, aromatic incense sticks, Indian curios and furniture, the latest music hits in Hindi or Tamil, freshly ground spices and posters depicting your favourite Hindu film star. Women from Chennai proffer Indian mangoes and various Indian knick-knacks, laid out on the pavement. The restaurants in the area offer all the well-known South Indian dishes, such as Masala Dosa (a kind of pancake with spicy vegetarian filling), Idli (steamed rice cakes) or as-much-as-you-can-eat meals, served on environment-friendly banana leaves. If in doubt about your future, you may consult an Indian soothsayer and his tireless little parrot - the latter will expertly pick out your prediction from a pack of cards… All things said, "Little India" comes across as a cleaner, more wholesome version of Madurai or Madras.[41]

Little India indeed offers the unique experience of all the goods and conveniences of a south Indian district within the city limits of Singapore in an environment perceived as safe and conducive for businesses to thrive. The variety of goods and services provided by the businesses within their intensively cramped premises provide the Singaporean Indian community with the vital link to the material culture of Indian life, such that the geographically dispersed community finds its common ground within its public space.

3.3.3.1 Cultural Practices

The public spaces here are the veritable stage of everyday Indian life, the space of social production. These public spaces are at times neither extensions of buildings, nor are they represented by urban spaces, but are created through the actions of everyday life and the symbolic life of Little India, i.e. through spatial practices (Fig. 3.14). Indian everyday life practices that are re-enacted include the buying and selling of Indian goods, the production and consumption of Indian foodstuff and cuisines, the daily use of southern Indian languages and dialects for communication, the form and rituals of Hindu and Indian-Muslim worship and the consumption of Indian movies, books and music, such that the social structures and form of south Indian ethnicity are re-produced in public space. The spatial practices that allow for the production and re-production of ethnicity and identity take place in public space, making Little India the symbolic realm of social representation in the Arendtian sense (1971), and thus the space of appearances.

Such a notion is most intensely reflected in the annual street procession of the religious festival, Thaipusam, the most important of Hindu festivals. Celebrated in

[41]R. Krack, *Discover Singapore's Little India*, 2001, CPA media, http://www.cpamedia.com/20011204_02. This website is no longer active.

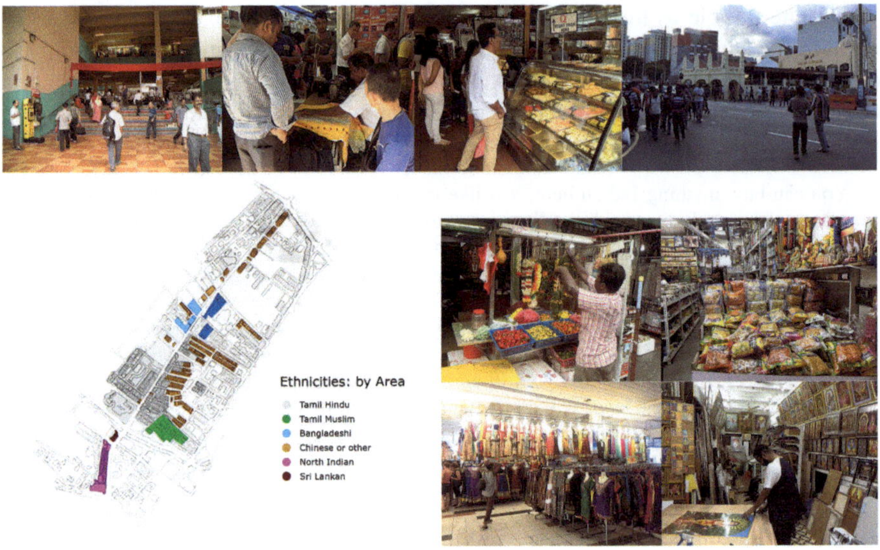

Fig. 3.14 Ethnic space

late January or early February, it celebrates the consecration of the deity, Lord Subramaniam, regarded as the all-powerful granter of wishes. Devotees take the opportunity to ask for a future favor, to fulfill a vow for a favor received, or to repent for past sins and wrongdoings. In a public display of their devotion, believers walk three kilometers from the Sri Srinivasa Perumal Temple at Serangoon Road to the Chettiar Hindu temple at River Valley Road, enduring some form of physical pain such as body and tongue-piercing or carrying a structure supported by sharp hooks skewering the human flesh, called a *kavadi*. Musicians, trance dancers, family and supporters of the devotees accompany the procession. The event is a huge spectacle with many onlookers, who range from other members of the Indian community, other Singaporeans, expatriates and tourists. The symbolic spectacle reinforcing the identity of the Indian community within the main street of Little India continues to establish the district as the public space of the community, so much so that it had also surfaced the notion of "insiders" and "outsiders" defined in relation to each other:

> Still, it was an extraordinary event. It's not simply tradition or ritual, like so many religious ceremonies I've seen. This was an intense demonstration of belief and devotion. It was a public event, attended by thousands, yet it appeared that every individual was alone at some point, even amidst the throngs, communicating with their god. Glimpses of these personal moments sometimes made Kevin and me uncomfortable, as though we were intruding, but these were, of course, the moments that stick with us.[42]

[42]This extract comes from a blog posting by G. Hughes, an expatriate worker in Singapore, 1998. See http://glennh.tripod.com/ROVWhat.htm.

3.3.4 *"Insiders" and "Outsiders"*

Siddique and PuruShotam (1982) had outlined two types of people inhabiting the space of Little India: the "insider clientele" and the "outsider clientele". The "insider clientele" are members of the large labor force—those who keep the businesses and eating houses running and who live in lodgings in the vicinity; "the outside clientele" is defined by those who buy daily and festive necessities—mainly local ethnic Indians, buyers of "Made in Japan" electronic goods and *sarees*, Indian tourists, and buyers of Indian exotica, including local non-Indians and non-Indian tourists (Siddique and PuruShotam 1982, pp. 85–94). However, it is noted that Siddique and PuruShotam's study was conducted in the early 1980s, when foreign workers from South India, Sri Lanka and Bangladesh were hardly seen in Singapore. The arrival of these workers in the 1990s and their affiliation with the district had surfaced a more complex understanding of what constituted "insiders" and "outsiders", as the phenomenon engenders not only a greater polarity of groups inhabiting or visiting Little India, but also leaves us with shifting and multiple subject positionings of these groups (Fig. 3.15).

For example, when we speak of "insiders", we may now not only think of the Indian residents, business people and workers who populate the quarter, but also the non-Indian residents, mainly the ethnic Chinese, who live in the public housing developments and run Chinese businesses like jewelry shops within the district. To this group, I would add the regular clientele who patronize the businesses and

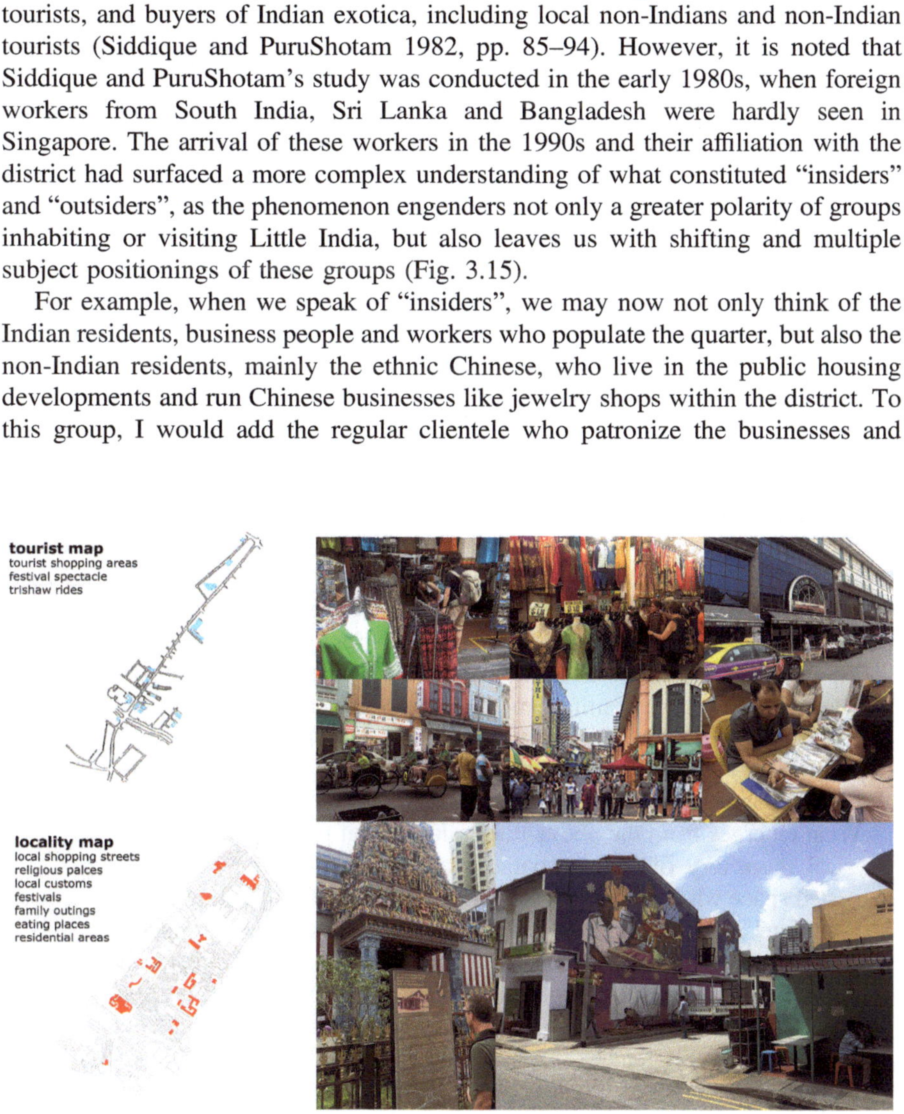

Fig. 3.15 Tourist/locality map

partake of the religious life of Little India, revolving round the temples and mosque there. Relative to non-Indian tourists, the Indian tourists who regularly shop at the district en route from other parts of the world to India, would consider themselves the "insiders" who possess the local knowledge in getting the best bargains and are regular customers to the businesses in the area. Singaporean non-Indian visitors in the district, in relation to non-local visitors to the area, would consider themselves "insiders", as they are familiar with their favorite banana-leaf fish head restaurant and shops there.

Even the foreign temporary workers who visit on late Sunday afternoons have strong affiliations to the area, as here, they speak in their native tongues, eat familiar foods and find some medium of entertainment in the city which they associate with 'home'. To them, the district is in many ways 'home away from home', where they could meet friends and socialize, get news from their villages and remit their hard-earned money to their families back in south Asia. Little India is the associative public space where familiar cultural and spatial practices re-create the common identity of 'Indian-ness'—even when these workers come from places as diverse as Madras, Bengal, Sri Lanka and Bangladesh.

On the other hand, while the non-Indian tourists are clearly the "outsiders" in many regards, they share the knowledge that Mustafa Center offers the cheapest electronic goods and products not limited to those favored by the Indians. The famous bargain basement department store was built by a local Indian entrepreneur, who started his business from a humble convenience store and made some shrewd real estate investments thereafter. Even the hordes of Chinese trishaw men who ferry groups of Japanese tourists around the quarter are such a familiar sight that it is not clear if they should be regarded as "outsiders".

While the ethnic Chinese population here consider themselves the "insiders" in their neighborhood, they remain outside of the social, cultural and spatial practices of the Indian community. The antagonisms that they experience with the congestion caused by the Sunday enclaves of south Asian foreign workers tend to reinforce their position as "outsiders". As such, the subject positioning of the various groups inhabiting public space continue to shift relative to each other, and in response to changes such as the increase in the numbers of foreign workers, gentrification resulting from the Conservation Plan and the lifting of rent control, growing tourism and the graying demographics of Little India.

What emerges from the above discussion is that public space here is not only a physical repository of Indian spatial practices and culture, but exists also as a contested space where different groups with sometimes conflicting interests co-exist and continue to define themselves through spatial practice. The subjectivities of these groups are socially constructed by everyday spatial practices within the context of Little India. The Chinese trishaw riders, while ferrying tourists round their familiar routes in Little India, are "insiders", but they become "outsiders" the moment they are no longer "tour-guides", as they remain outside of the social network of the ethnic quarter. The most poignant change in the wake of the riots is the voluntary exclusion of some migrant workers, traditionally considered "insiders" to Little India, choosing to become "outsiders" to the space. As Little India is re-configured along the lines of public safety, many migrant laborers feel unwelcome by the new policies regulating public space (which will be discussed later in this chapter). Many Indian and Bangladeshi migrant laborers have begun seeking out and congregating in alternative weekend hangout spaces, as recounted by 29 year old Satishkumar Katakam, a migrant electrician from India who first came to Singapore in 2007, and returned again post-riots in 2014. What was once a space of affirmation for migrant workers who have a stake in Little India both culturally and socially, has now become a space of antithesis, where their role as "outsiders" is averred.

> Because whatever that happened, other people doing like that, so there is some fear in everyone's hearts to come to Little India... So that's why my friends and me also, one day also I don't want to come to Little India... I just send money. I don't want to stay here also... lot of workers they are very worried to come to Little India. Before that riot, there are a lot of people every Sunday we come here we hang somewhere. Then we had freedom. Now sometimes I call my friend, 'how you want to come meet me or not?' They say 'why we need to go Little India?' Still they are very scared to come to Little India. They say okay we just (go to) some other place, Jurong East or East Coast Park or Marina Bay. We go some other place lah, better.[43]

3.3.4.1 Global Entities

Migrant laborers from south Asia interact socially and spatially with the people and context of Little India. These social and spatial processes are mutually defining, creating a continually evolving identity for the district. Here, any attempt to define the 'global'—in this instance, the lower end of the transnational flow of human labor—as universalizing and the 'local' being sites of resistance to these forces of change completely breaks down. The 'global' in this instance, being often originating from the poor, rural context of south Asia and ingrained with its own cultural habits and spatial practices, interacts with the 'local'—in this case, the ethnic Indians of Singapore who are socialized into living in a rapidly modernizing city with a heterogeneous population. In many ways, the 'global' entities in the public space of Little India carry out spatial practices that reinforce the notion of Indian ethnicity rather than have any universalizing effect.

In addition, the quarter also functions as a portal linking south Asians here with Indian communities in other parts of the world. Here, Indian workers can get news from home, make long distance phone calls in the numerous phone booths installed here back to their home towns, remit money and send letters to their families as well as purchase commodities from their home countries. That Little India is also a draw to tourists from India who come on shopping trips to purchase electronic goods and other high quality products made for the Indian market but imported from places like Japan (a niche Mustafa Center owes its success to), speaks for its role as a

[43]This comes from an excerpt from an interview with Satishkumar Katakam done in November 2015.

global portal which Indians from all economic classes tap into. The Indian-owned businesses in Little India provide a familiar environment for the transaction of foreign products for these tourists. The overlapping interaction of ethnicized and globalized space creates a unique sort of public space serving as a common ground for the diverse strata of the Indian Diaspora.

3.3.5 *Space of Exchange*

The public space of Little India offers itself as a space of exchange—of commodities and goods, friendship, information and services. These take many forms, which I will discuss through the framing of notions of "front" and "back" activities. "Front" activities are those carried out in the bright open public spaces—state-sanctioned activities that contribute to an idealized notion of the ethnic quarter. "Back" activities include the chaotic, sometimes clandestine or merely less visible activities played out by those who possess intimate local knowledge of the area, activities bordering on the edge of legitimacy, or those fleshing out the 'dirty realism' of the lesser seen face of Little India but contributing no less to the holistic notion of its public space.

3.3.6 *"Fronts"*

The notion of "front" spaces, in many ways, corresponds to what Henri Lefebvre, in his three moments of space, has described as "representations of space" or "perceived space" (1991, pp. 68–168). These consist of a system of prevailing signs and codes that outsiders can relate to and associate with a particular social group, environment or mode of production. In this context, the frontal spaces are situated in public space, which are constituted through associative spatial practices and cultural artifacts of the south Indian community, and not necessarily through urban morphology, setting the area apart from other public spaces in Singapore.

One observation that I have made is that although men and women are found in the most visible public spaces of Little India, women are ubiquitously absent or not overtly present in the "back" spaces. For example, ethnic Indian women are found most often shopping along the five-foot ways of shophouses, buying *sarees* and fabrics, and foodstuff; or they may be vending religious artifacts, Indian sweets and confectionary, garlands and flowers, or fortune-tellers. Women in public spaces are often with their families on the way to restaurants, accompanying children, or are chaperones to younger women. The Indian male-female relationship entails "a greater conformity to maintaining physical distance" (Siddique and PuruShotam 1990, p. 82) and it may be presumed that interactions with males other than family members are kept to a minimum through their lack of presence in spaces more often patronized by men, such as at the outdoor beer garden at Buffalo Street and spaces

Fig. 3.16 Gender + space/subversive spaces

of dubious reputation such as the Desker Road area. In other words, the public spaces that women inhabit coincide in many ways with the main streets and five-foot ways, which are often also the spaces of "walks" that tourist maps of the ethnic quarter prescribe (Fig. 3.16).

This leads us to the other obvious "front" spaces—those proffered in tourist maps of the area. These are often in the form of carefully constructed routes—such as walking tours or, as preferred by East Asian tourists, tours via trishaws—a three-wheeled open pedi-cab driven by a rider—an option that keeps the engagement of the district at arm's length. The tourist map of Little India presents a prescribed sequence of places and events that unfold predictably to reveal "endless dining options, ethnic shopping and cultural treasures this lively district has to offer".[44]

The images and sensory perceptions here are packaged to be consumed as vignettes: the covered shopping streets of Little India Arcade, the sounds and prayer rituals at the three Hindu temples on Serangoon Road, the smells and scents emanating from the Indian restaurants and spice merchants, the colorful *sarees* and fabrics on display, and the sensory assault of the Deepavali light-up for a month every year. The inhabitants and culture of the quarter are packaged as a homogenous entity ("the Indian ethnic quarter of Singapore") and its public space a stage for the re-enactment of cultural practices and customs. Thus the tourist perception

[44]This phrase comes from a 2002 brochure by the Singapore Tourism Board, *Little India: Ethnic Quarter*.

of the public space is blind to areas and nodes outside of the prescribed route, and remains two-dimensional, not penetrating beyond the bounded bubble of tourism.

The representative "frontal" public spaces that produce the image of the Indian ethnic quarter are highlighted in this case, namely the concentration of the south Indian religious buildings of the Hindus, Muslims and Buddhists, the visible and performed rituals of Indian religion (e.g. prayer rituals), culture (e.g., embellishments of the body) and those of the trades (especially petty tradesmen and vendors occupying the five-foot ways), extending even to the sensory expectations of a gamut of colors, some degree of disorderliness and an assault of the olfactory senses. The experience is somewhat exemplified by this blog entry:

> It felt somewhat surreal, as if I was in another country … as I wandered through the smaller roads - such as Dunlop Street, Campbell Lane, Dickson Road - that were linked to Serangoon Road. I suppose to get a more 'complete' feel of what Little India is like, one has got to explore those roads away from the main road. The roads were lined with quaint shop-houses, not unlike those in Chinatown actually, complete with the traditional five-foot way. Lots of colour too - sari cloths, fresh flowers, fresh vegetables… & gold (courtesy of the numerous Indian jewellery shops)! The atmosphere amidst the smaller roads was rather laid-back, & that's partly why it made me feel like I was not in Singapore (when more often than not, most people are just rushing around in the daytime, and hardly stopping to take a breather). Oh, and guess what I saw at the Little India Arcade? "Barbie in India" collection! I think it's authentic (did some snooping around on the Internet). Cool.[45]

"Backs"

In counterpoint to the tourist map, I introduce the notion of the "locality" nested within the concept of the "back" spaces. The notion of "back" spaces or those generally not visible or obvious to an outsider, is similar to Lefebvre's notion of "representational space" or "lived space"—these spaces are directly experienced by users and inhabitants in their everyday life and become, through their symbolic and social significance, the imaginative space of the community. The locality in this context is constructed from a common pre-urban and urban history—an urban village that grew from similar circumstances involving migration and subsequent entrepreneurship, it is geographically (due to its growth off Serangoon Road) distinct from other parts of the city.

The transition from tourist space to locality is not one in which a distinct threshold is crossed, but one which taps into a network of public spaces—a "field" of spaces of everyday life rather than a prescribed sequence of spaces. Many of the spaces on the tourist map do in fact overlap with spaces of the locality—and are embedded in the locality map via dense networks of connected spaces rather than as isolated vignettes. Such spaces include mosque and temple grounds and their vicinities; the five-foot ways which often serve as social spaces, especially at the corner locations of coffeeshops; the street food and vegetable markets along Dunlop Street; the ground floor common areas of public housing blocks; spaces around the local church and community center; back lane coffee shops and beer gardens; the

[45]This excerpt is from a blog posting by Angelina Hue, 23, Singaporean non-Indian, trainee tourist guide. The blog posting is no longer active.

wet market and hawker center within the Tekka Center, which serves the daily needs of the local community regardless of ethnic group; and especially the areas around the Tekka Center, which often serve as a rendezvous points for meeting with friends and acquaintances.

What makes these spaces part of the locality is their embeddedness within the daily routines and spatial practice of the community. These spaces are the incidental spaces where the everyday life of Little India is continually re-enacted and where spatial practices within these public spaces create a sense of community. The mental maps of the locality exist within the minds of the local inhabitants and regular clientele of the businesses here, and sets apart the "insiders" from the "outsiders" discussed earlier in this chapter. The mental maps of the public spaces of the locality are also different relative to the subject positioning of the perceiver, which is influenced by aspects such as gender, age, occupation and even specifically the part of south Asia from which the subject hails, as well as the individual experience and symbolic importance associated with these spaces. The overlapping networks of the public spaces of the locality allow for routes to cross and meetings, whether routine or serendipitous, to occur.

The locality, however, is not a homogenous entity. Tamilians constitute a majority of the Indians in Singapore, but there are also many discrete ethnic groups such as the Sikhs, Gujaratis, Chettiars, Bengalis and Parsees, not to forget that other south Asians such as the Bangladeshis and Sri Lankans (Ceylonese) are here as transient foreign workers (Fig. 3.17). The diversity of the community is reflected not just through the variety of religious buildings that can be found in Little India, such as the Hindu, Buddhists, and Sikh Temples, Islamic mosques and Christian churches, but also through the location and concentration of businesses and food establishments serving the disparate groups. In other words, the distribution of urban amenities throughout the quarter is shaped by historical associations of certain groups with certain micro-districts, creating an uneven cultural field that influences the location and affiliation of subsequent groups in the area.

For example, the areas north of Serangoon such as the location of cattle pens (e.g. at Buffalo Lane and Kerbau Lane) were traditionally Hindu domains, whereas the areas south of Serangoon Road, especially near the Abdul Gaffoor Mosque, were Indian-Muslim areas. The Tamil-Hindu shops such as jewelry and textile businesses and restaurants are concentrated around the main Serangoon Road and the Race Course Road areas, while the Tamil-Muslim businesses and restaurants are at the Clive Street area south of Serangoon Road. Interspersed along Serangoon Road, one finds the Chinese-run jewelry shops, while Chinese hardware and sundry shops can be found on Dunlop Street, Desker Road and Rowell Road. The insertion of public housing on Serangoon Road and at Rowell Lane injects into the area, a resident ethnic Chinese population who find their groceries and food stalls at Tekka Center and also patronize Chinese-owned coffeeshops in the neighborhood. The Tekka Center is a commercial node of the area and attracts a mixed clientele, while tourists in general are drawn to the gentrified Little India Arcade. Indian tourists in particular are drawn to the Mustafa Center on Syed Alwi Road and to Serangoon Plaza.

Fig. 3.17 Macro-realms distribution in Little India

Within the field of localities, an even more deeply embedded field of "public" spaces exists, situated in the ambiguous space between the public and the private, the legitimate and illegitimate. These are the "subversive spaces", where users and inhabitants of these spaces have a strong wish to remain clandestine, and these spaces exist only through the shared local knowledge of the area (Fig. 3.17). Outsiders who are not privy to this knowledge would often find themselves excluded from these spaces. Foucault's rendition of heterotopic spaces (1984) provides the closest description of such spaces, a kind of counter-site through which real sites are "simultaneously represented, contested and inverted. Places of this kind are outside of all places, even though it may be possible to indicate their location in reality". However, what comes closer would be the category of heterotopia which Foucault terms "heterotopias of deviation: those in which individuals whose behavior is deviant in relation to the required mean or norm are placed" (ibid.)

Although the Desker Road and Rowell Road areas are Designated Red-Light Areas (DRAs) in the planning terms of the URA, they are considered a necessary evil rather than sanctioned sites of leisure production. The subversive nature of such "public space", especially concentrated at the back lanes of Desker Road and Rowell Road, are signified by the ominous signposts of 'bouncers' at the entrances to the back lanes which lurk under big umbrellas set up on a pile of tires, barely hiding the fact that they are on the look-out for some sign of 'danger'. Although I have explained that prostitution is not illegal here, whole paraphernalia of illegal activities accompany the business, such as the peddling of adult printed materials, videos and sex toys, soliciting and pimping. As such, the bouncers who keep watch at the entrances to the lurid back lanes play a role in keeping the notion of heterotopia alive, as "(h)eterotopias always presuppose a system of opening and closing that both isolates them and makes them penetrable" (ibid.)

In general, the heterotopic site is not freely accessible like a public place. Either the entry is compulsory—as in the case of entering a barracks or a prison—or else the individual has to submit to rites and purifications. To get in, "one must have certain permission and make certain gestures" (ibid.) Unintentional visitors who are not privy to the signs and signifiers of the nature of this space would often be confused, in that "(t)here are others, on the contrary, that seem to be pure and simple openings, but that generally hide curious exclusions. Everyone can enter into the heterotopic sites, but in fact that is only an illusion—we think we enter where we are, by the very fact that we enter, excluded" (ibid.). This phenomena is illustrated by the following account:

> I parked in Desker Road, a well-known haunt for prostitutes and transvestites. This wasn't on purpose, most parking there is a challenge for me (if you are bored and would like a parking challenge, try the old Mustafa carpark) - this road still had lots I could just drive into.
>
> Desker has been quiet for some time (not that I really know :)) but I realised lately a bazaar of sorts has mushroomed. You get fresh fruits and vegetables and an assortment of food stalls catered to native Bangladeshi and Indian workers. Many people find this area scary but I think that's a perception - look beneath the surface and you get good fresh produce, interesting food and stuff e.g. ear candling, eyebrow threading, great kulfi and spices.
>
> Turning into one of the small lanes, I spotted the sign "Ayurvedic Massage". I tracked up the steps and opened a suspicious door - only to be greeted by a dog. Having had several strange massages before, I persevered in search of the Ayurvedic grail...the receptionist was a man (Chinese). Just tired of walking, I signed up for a 30-minute head and shoulder massage. The therapist was Chinese too! This lady looked at me most strangely. Later I found out from her (in the course of my massage, this place was mostly frequented by men – gee, explains why the toilet only had a male sign.[46]

[46]This excerpt was from the blog of the Spoke, female Singaporean. The website is no longer active.

The spatial setting for the sex trade and its dubious activities in the area is mainly centered round the back lane between Desker Road and Rowell Road. The two streets are lined by two-storey shophouses of which the fronts face the street. But while legitimate businesses occupy many of these units, intermittently found between these are units that are devoid of signboards on the front and have doors and windows permanently shut, betraying no signs or signifiers of their usage except for the occasional red Chinese lanterns. While these units face the street, the units at the ends of the streets are turned ninety degrees such that a kind of entry to the back lane area is created, flanked by the perpendicular city blocks. Coffeeshops, usually found at the corners of the street units, generally spill out into the back lanes. They often have the makings of a tawdry beer garden, with colorful umbrellas and open-air seating benches. These form a sort of buffer at the entrance to the back lanes, in addition to the 'bouncers' who keep watch over these entrances. Many of the clientele of the DRA at Desker Road are south Asian foreign workers and older Singaporean Chinese men, who, despite having little in common, share this subversive "public space" for the same reason. The following is a description from a tourist guide of a specialized nature:

> This is the classic low-end red light district. If you don't know where to go you might see no action, because it all takes place in the back alley between Desker Road and Rowell Road. Here you will find a teaming mass of humanity milling along the alley, looking in at all the doorways. Inside the doorways you will see one to six women sitting around looking bored and contemptuous in a bare and uncomfortable room…[47]

The back lane functions as real fronts for the nondescript units from the street-side. The walls on the back lane display the red hand-painted unit numbers. The doorways lining the lane often have a temporarily slung-out canopy over to shield the entrances not only from the sun and rain, but also from views from the surrounding high-rise residential public housing. The potential clientele of the sex trade cruise along the back lane in groups or alone at all times of the day, peering through each open doorway in turn. The back lane is punctuated by temporary folding tables displaying the related paraphernalia of the businesses, such as contraband pornographic compact disks and sex toys—these are ready to be folded into plastic sheets and grabbed in case of police raids. There is a conspicuous absence of women in the back lane spaces—glimpses of women are only possible through the unmarked doorways leading into the back areas of the shophouses. Here clearly, women are objects of the men's gaze.

The heterotopic subversive space of the back lane is preserved by unmarked entrances, the inversion of the shophouses along the street, and the signs and signifiers of the space which only those privy to the local knowledge of the area can decipher. Other kinds of subversive activities such as illegal betting and gambling also take place in the back lanes. The spatial nature of the back lane is defined by its enclosure, the degree of visual control, the thresholds of interiority that one has to

[47]This excerpt is found on the 1995 World Sex Guide Website: Report on Singapore. http://www.worldsexguide.org/sp_travel.txt.html.

penetrate from public to private, the gender-based rules of operation, and the selectivity of the participants who have accessibility. In other words, the "other place-ness" of the space allows for deviant spatial practice not normally sanctioned in "front" public spaces, and isolates this particular slice of Little India from the other public spaces:

> The route took me along a back-alley between two roads, lined by walls with entrances, which may have been back porches that were covered over. The whole area was dark, smelly, menacing. Lines of men walked slowly by–I didn't see anyone actually enter, or even take a close look at the women. The women, three or four per unit, sat around silently, looking hot and bored.
>
> Little shops along the way sold cure-all lotions, virility oil, sex aids. Stalls sold videos and trinkets. Somewhere around here were supposed to be the "temple pythons", curled up in cages and guarding money that superstitious men toss to them, but I didn't spot them.
>
> … (an acquaintance) took me around some of the more dubious streets in the neighborhood. On one of them, facing a huge apartment block, ghostly figures emerged from the darkness as we passed by, to stand in the dim shadows of the porches. Their thin pale faces were just visible, their heavily made-up lips and eyes emphasizing their ghostliness–when they pursed their lips in a sad imitation of a kiss they reminded me of the painting "The Scream". Behind this street was a narrower one, full of mak nyahs, healthier looking and fashionably, if skimpily, dressed (for a woman, that is). Many, with slim figures and hormone-induced breasts, could easily have passed as women. They crowded around us as we tried to pass through, attracting even more from nearby buildings until the whole centre of the street was full of them and I began to fear for our safety.[48]

3.3.7 Spaces of Boundedness

Due to the built-up urban fabric of Little India, what constitutes "public space" has to be intensely used, or even claimed from space not intended for public use. The contestations of space use arising from the tight urban configurations often result in public spaces which are bounded in some form. These public spaces are either spatially-bounded, i.e. its use is defined within a physical or imagined limits, or temporally-bounded, i.e. its use is defined within a stipulated time-limit, or perhaps both. One of the most commonly occurring extensions of space is the pushing of canopies, temporary tables and other paraphernalia onto the street space to claim part of the street as a marketplace, forcing pedestrians to venture onto the street instead of using the verandah five-foot way. This practice is prevalent on Dunlop Street, where a curb-side fruit and vegetable market causes the street to be almost off-limits to vehicles on weekend mornings, and also on Serangoon Road, where the congestion is caused by shop owners claiming the five-foot ways for the display

---

[48]This excerpt is taken from an internet posting by Dean, an Irish tourist to Singapore. The website is no longer active.

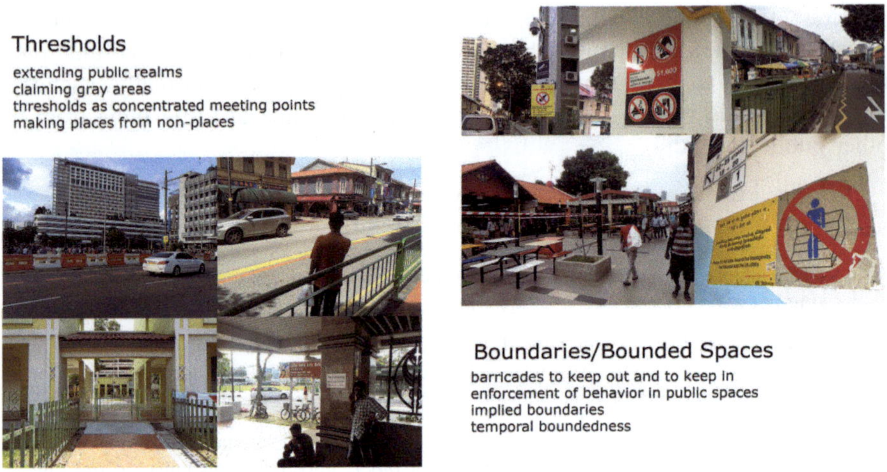

Fig. 3.18 Thresholds/bounded spaces

of goods, and also to deter "undesirable" clientele[49] from hanging around outside the shop if they had no intention to make a purchase. Furthermore, the state's reaction to the riots, subsequent alcohol policy, and its re-ordering of public space in Little India underscore how the legacy of the riots has decidedly affected the boundedness of public space in Little India today.

3.3.8 "Thresholds"

These spatial practices unequivocally cause pedestrians to have to use the street as thoroughfare, creating tension between the motorists and pedestrians over the use of space. This spatial culture of extending the limits of the public realm creates what I would term "threshold" public spaces—these spaces exist in a temporary spatial condition in a grey zone that has been spatially or temporarily defined as public space. In the same vein, the extension of canopies and umbrellas by corner coffeeshops onto the back alleys to claim extra space is such a common spatial practice that it has become a type of urban pattern to be found at the entrances to the back lanes here. These create a kind of threshold public space of the otherwise utilitarian back lanes (Desker Road aside). The use of plastic canopies as extensions of space and as a space-claiming device becomes also an established spatial practice, a kind of urban signifier read and understood by the inhabitants and frequent visitors of the area (Fig. 3.18).

[49]These are usually the foreign workers who throng Serangoon Road every Sunday evening.

The practice claims for use areas previously not well defined as either public or private space, i.e. spaces previously used as somewhere to pass through or in transition from one space to another become now public space of sorts. These peripheral or threshold spaces, due to their claimed or "stolen" nature, tend to be tight and the activities in there concentrated. As such, these are also intensely used social spaces, where people are brought into congested zones or into close proximities, increasing chances of social contact or "friction".

Other kinds of threshold spaces include the foreground of the (south) Indian Temples, such as the Sri Veerama Kaliammam Temple, which is often a very small space; the sidewalk outside the temple, the perimeter space surrounding the temple, and even spill-out space onto the streets. Women vendors selling flower garlands, the continuous stream of people moving in and out of the temple, the practice of removing and temporarily depositing footwear in the space outside the temple and the curious tourist armed with their cameras are all complicit in the congestion of the temple threshold spaces. The space immediately outside of the Tekka Center acts similarly as its threshold space—with people waiting to meet with friends, making phone calls, watching passersby or hawking wares as petty vendors. Unlike at the temples however, the public here comprises more of a mix than of any particular group. The nooks and crannies and the semi-covered space offered by the building edge form an ideal perimeter with a variety of spaces in which people can idle without having to have a real reason to be there.

One non-physical "threshold" that was implemented after the riot has been the new alcohol restrictions in Little India, which has been designated as a Liquor Control Zone. Little India, along with Singapore's red light district, Geylang, are designated Liquor Control Zones as they are places where "there is higher risk of public disorder associated with excessive drinking" (Straits Times 2015). Since April 2016, public drinking has been banned in Little India from 7 am on Saturday to 7 am Monday, and 7 am on the eve of public holidays to 7 am on the day after the holiday. Furthermore, on top of the daily island-wide ban of alcohol sales from 10:30 pm, shops in Little India are not allowed to sell takeaway alcohol from 7 pm on weekends, eve of public holidays, and public holidays. These new regulations fall under the Liquor Control (Supply and Consumption) Act, and aims to "curb public drinking to 'minimize public disorder and disamenities'" (ibid.). Failing to comply with the regulations in a Liquor Control Zone would carry a penalty of one and a half times of the $1000 fine for consumption of alcohol in a public place.[50]

It is almost impossible for individuals to claim they are oblivious to the law as they are constantly reminded of the "thresholds" to alcohol consumption. Such reminders manifest in the form of prominent signs, the threat of punishment and the omnipresent watchful eye of surveillance cameras—twenty-six additional CCTVs were installed after the riots (Straits Times 2013)—and enforcement officers. While

[50]See Liquor Control (Supply and Consumption) Bill. https://www.parliament.gov.sg/sites/default/files/Liquor%20Control%20(Supply%20and%20Consumption)%20Bill%201-2015.pdf.

these "thresholds" are an eyesore to some, to others, they are a passive form of structural violence against the well-being of migrant workers in Little India.

> "If foreign country people (tourists) see this sign, what will they think? Is it wrong, drinking? The workers they are working one week of hard work, they just meet up their friends on the weekend. They want to relax with friends, but even this also they are putting like that. It's not good, it's not good. You know, in the worker mind he (is) thinking: 'we are working very hard, and he is making rules in Little India. This is our place, and he is making rules in Little India.'...

> ... don't treat the people like everyone is the same. Not everybody is a criminal. It's happened, all those stories and histories. If one Indian is a problem, don't treat other Indians as the same. Every country there is a different kind of people. One country there is no all good people. Every country has a different kind of people. So, you can do whatever you need, you can take actions to make sure this situation doesn't repeat again. But you don't treat the workers as like criminals" (Katakam 2015) (Fig. 3.19).

3.3.9 "Sunday Enclaves"

A more complex form of bounded spaces, both in the spatial and temporal sense, exists within the phenomenon which I will describe here as the Little India "Sunday Enclaves". Every Sunday evening, an estimated 30,000 south Asian foreign workers, mainly from south India, Sri Lanka and Bangladesh, converge onto the public spaces of Little India. Due to the sheer volume of the crowd, they literally transform the landscape and bring with them new forms of spatial practices. These spatial practices in turn lend a different take to how public space may be used, or what in fact constitutes the notion of public space.

As I have mentioned earlier in this chapter, this phenomenon was unknown before the change of labor laws in the late 1980s to enable the employment of

Fig. 3.19 CCTV cameras and Signages along Little India. *Source* Author's collection

temporary south Indian workers to alleviate the shortage of workers in some economic sectors, in particular, the construction, manufacturing and shipbuilding industry. Foreign workers in Singapore are regulated by immigration and labor laws, and are not supported by government welfare agencies. They do not participate in the Central Provident Fund (CPF), and have no rights to permanent residence in Singapore. As such, these transient workers have little legitimacy in Singapore and have virtually no stake in the country, except to save their wages to take back home after their short stints here.

On a regular weekday, these workers remain as invisible cogs behind the engine of the growing economy, kept hard at work by day and contained within worker dormitories by night. But come Sunday evening, their only day off, they throng the streets and spaces here, beginning from about five in the evening. These temporary enclaves in public space create a realm of temporally bounded spatial practice that is in sharp contrast with the everyday use of public space. The temporary appropriation of all kinds of open spaces, including those not normally considered public space, imposes a system of spatial practice over the existing spaces, engendering a whole new topography of public spaces. The transformed public space of Little India presents itself as a site of spatial improvisation, but also a site of manifestation of already existing forces within public space. New meanings and spatial practices are borne out of the intense appropriation of space, which yield lingering influences and traces within the public space outside of its temporal boundaries (Fig. 3.20).

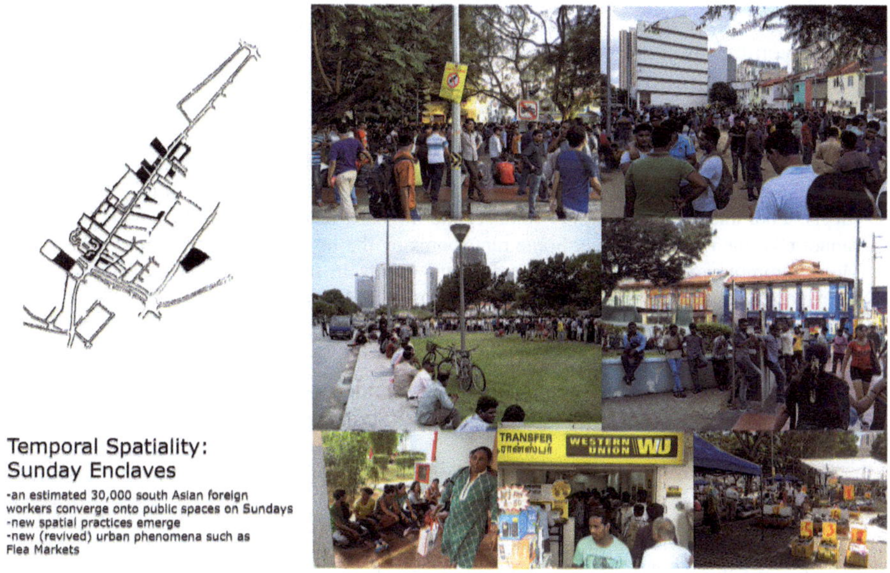

Fig. 3.20 Temporal spatiality: Sunday enclaves

Many of the transient workers are Tamils who come from south India, and Bengalis from India and Bangladesh.[51] Activities for the transients include buying provisions, remitting wages back to their villages, sending letters and parcels home, meeting with fellow countrymen from other worksites, exchanging news and gossip from home, using the myriad phone booths installed in the area to call home, going to the cinema or engaging prostitutes. Eating is not so commonly indulged in, as most would have had a meal at their dormitories before being deposited in the area by their company buses.[52]

The Sunday enclaves of workers occupy almost every available open space here. However, what at first seems like utter chaos, on a closer look, reveals spatial logic and forms of social structure. While the crowds may spill out onto the streets, most of the social interaction takes place at street corners or in open spaces and car parks, back lanes, or even in the five-foot ways—in other words, the crowd splits into flows and nodes. The nodal points for gathering are not totally random, as different ethnic groups within the heterogeneous makeup of these workers are concentrated in different spaces.

While the tendency is to congregate according to native tongues, religious and cultural affiliations, the actual spatial proximity of these spaces to preferred goods, services and food outlets also play a determining role.[53] For example, Tamils congregate around the perimeter walls of and in the open car parks adjacent to the Hindu temples, while Bengali Muslims can be found in the vicinities of the mosques and near the Indian-Muslim owned Mustafa Center, and outside Serangoon Plaza. It may be ironic, but the one place in which there is a genuine mix of all ethnic groups and ages, including Singaporean middle-aged to elderly men, is within the Desker DRA. The absence of women and children in public spaces on Sunday evenings is just as stark as the male dominated Sunday enclaves. There is also a vibrant homosexual scene at some parts of the district, as witnessed by this blog posted on a gay website:

> There is a vibrant cruising culture for young Indians, mostly unmarried, in Singapore. Around Serangoon Road from Teka market to Mustafa's, are bus stops, facilities, quiet side streets, a bar and boutiques where contacts and intimacy take place. On the whole, Indians cannot take their friends home. Some rent rooms by the hour in shabby Serangoon hotels. More common are public liaisons and group events in dark corners, especially where redevelopment is taking place…The plaza in front of Mustafa's always looks like a disorganized South Asian summit with Hindi, Sinhala, Urdu, Persian, Bangla, Nepali, Burmese and Mongolian speakers shouting to be heard.
>
> Within this chaos, there is heavy cruising. Some of the men are rough trade. Most of this happens in late evening with the focus gradually shifting to 24-hour kopitiams in nearby

---

[51]T.C.S. Yeo's study of *Socioscapes in a Global City: Singapore's Little India* (1999) notes that 60 % of the workers hailed from Bangladesh.

[52]While it is true that many workers arrive in company buses, many do take public transportation like buses and the MRT.

[53]Many south Indian workers are Hindu vegetarians, while many of the Bengalis from Bangladesh are Muslims.

> lanes. Not many local Indians dare to be seen in the Plaza. The married ones will venture to a favorite bar, but they get their contacts on the phone and through friends, sometimes even the family priest. The transvestites behind Desker Road always attract a large, non-local, Indian presence. Just look for the gawking crowd of lungi-wallahs near the cigarette stall, and you're there. Here, there is little stigma to be seen ogling transvestites because they are "honorary women" as opposed to gundus or kothis.[54]

The foreign workers tend to congregate in dense groups within open spaces like car parks and green fields, often sitting shoulder to shoulder. Some would sit while others would stand around in closed clusters, but their activities are generally passive, such as chatting, watching, or napping, and occasionally soliciting or indulging in voyeurism on the Desker Road back lanes. The improvised spatial practice of these workers, who occupy spaces such as along roadside curbs, lining drains, along footpaths, sitting on railings and along perimeter walls of buildings as well as sleeping in public "break" the rules of what is regarded as "correct" behavior in Singaporean public space by common consensus.

As for the nature of complaints received, these pertain to foreign workers congregating and congesting the common areas and footpaths of the residential estates, vagrancy, noise pollution, traffic congestion, littering, urinating in public, etc. Other reports include the occasional reports of fights and crimes (Yeo 1999).

The large numbers of the congregations and the "foreign-ness" of these silent groups of virile males have also conjured up notions of "fear of crime" and of resentment towards these groups especially by women, non-Indian residents,[55] business-owners who do not benefit from their presence, motorists who have to drive through congested streets, and local visitors. There was the case of the Chinese Singaporean MP who was rebuked for his remark in Parliament that it got dark very early in Little India, referring to the masses of Indian and Bangladeshi workers who congregate there on Sunday evenings.

While loitering in public space is an offence in Singapore, the role of the state in dealing with the phenomenon—which it had in a way brought about by importing foreign labor—is one of a conflicting nature. The stance taken by public authority regarding the problem of crowding in Little India had been one of pragmatism and improvisation. It is a well-known fact that the Singapore government would not hesitate to clamp down hard on any incidences of breaking the law, or of inappropriate behavior in public space. However, it would hardly be pragmatic to throw half the foreign worker population here in jail for jaywalking or loitering. As reiterated by the Police,

> ...The foreign workers, coming mainly from the undeveloped countries of the Indian sub-continent, are not accustomed to our traffic scheme and system, and invariably jaywalk or spill onto the main roads, posing danger not only to themselves but also to motorists as

[54]Excerpt taken from Utopia Asia (http://www.utopia-asia.com/tipssing.htm), 22 July 1998.

[55]From T.C.S. Yeo's field interviews of foreign workers in Little India, ethnic Indian residents of Little India tend to be more sympathetic to the plight of these workers than non-Indian residents, whose lives are inconvenienced by the physical congestion caused by the crowding and the intangible fear of potential criminal activities.

> well…We monitor the congregation of the foreign workers closely to ensure that the enclaves created along ethnic lines do not erupt into communal clashes triggered off by events and conflicts back home (Yeo 1999).

Instead, the state's stance had been one of controlling and apportioning rights to space for the parties involved in the conflicting use of space. Over time, the way authorities dealt with the problem of congestion here had even taken on rather creative directions. The state is present in the form of temporary traffic control towers set up every Sunday evening and police officers in luminous vests who ensure that foreign workers who cross the street without heed of traffic signals make it across safely, with some compromise to the smooth flow of traffic. While the men in blue tend to leave the workers who congregate in open spaces alone, the workers are mostly kept off pedestrian footpaths, such as those around Tekka Center, so as not to obstruct the public right of way. The highly visible presence of the state in public space in the form of police officers and temporary police posts also tend to assure non-Indian residents of Little India that things are under control.

3.3.9.1 Barricades

The phenomenon of the foreign worker enclaves also bring to light the purposes of barricades and fences which have been put up either specifically for the Sundays or are semi-permanent. Barricades surrounding the common areas of public housing developments are quite unique to this area. Railings line and define walkways outside Tekka Center and along the road, in some cases clearly dividing publicly owned land from privately-owned land. Some green reserved land parcels are also fenced to prevent the foreign workers from using these spaces to congregate on weekends. Some stretches of five-foot ways are barricaded physically with gates and grilles, while some businesses fill these up with goods and displays to prevent them from filling up with foreign workers, who generally have little money to purchase goods. These barricades look strangely incongruous on weekdays, but perform the passive controlling function on Sundays. Even more barricades have sprouted up along Little India in an effort to discourage jaywalking and create orderliness in Little India, post-2013 riots. They are the physical manifestation of the bounded spaces (Fig. 3.21).

3.3.9.2 Sunday Market and Spaces of Informality

> Last year, the Police, working with nearby community organizations, set out to provide alternative venues for these foreign workers to congregate on their rest day. The vacant plots of land at Weld Road proved to be ideal for this purpose. It would provide some relief to the residents of Zhujiao and Rowell Estates by drawing some foreign workers away from their overcrowded neighbourhoods. The idea was first tried out in November last year and it had grown into a Sunday Market at Weld Road.

Fig. 3.21 Plastic safety barriers and metal railings. *Source* Author's collection

> When the Sunday Market was started last year, it attracted only small crowds of about 150 foreign workers each Sunday. The crowds have since grown to about 1000 on Sunday. We expect the crowd to increase as news spread by word-of- mouth, and as the facilities and attractions are further improved.[56]

The Sunday Market in Weld Road, held in an open space just across Jalan Besar Road (one of the bounding roads of what I have defined as Little India), was officially launched by the government to relieve the congestion in the area, and perhaps to alleviate the critical mass of the Sunday enclaves. The phenomenon of the Sunday flea market around this location, originally called the Sungei Road Market or Thieves' Market, had itself been outlawed since the 1980s, when the overcrowded shophouses in the area had been demolished. The market in fact had its roots in the 1930s, as an outlet for British army cast-offs—"...parachutes, knapsacks, billycans, raincoats and boots all found their way to Sungei Road in search of a new owner" (Gopalkrishnan and Perera 1983, p. 99).

The eventual resurrection of the market occurred around 1995[57] at the same location, which now consists of an odd landscape—a narrow street without street signs, lined only by the ghosts of long demolished shophouses. The market arose out of a series of serendipitous phenomena which seemed to sustain its persistence, and it has continued to exist despite its illegality. Stalls at the Thieves' Market, now renamed the Weld Road Sunday Market, consist of makeshift folding tables, tents, simply rolled out canvases on the ground that boast oddments, old electrical appliances, toys, car accessories, used clothing, and a whole assortment of indescribable thingamajigs. Public space was, in fact, "stolen" once again—by the Weld Road Sunday market.

The prominence of the Sunday market, its undeniable weekly life and the throngs of foreign workers who are clientele to the market would not have escaped

---

[56]This excerpt is taken from a speech by Lee Boon Yang, Minister for Manpower, at the launch of the Sunday Market, October 1999.

[57]An article by Dinesh Naidu. "Sungei Road: a market on the margins" in *Singapore Architect* (1999) covered the phenomenon in detail.

the authorities in its informal revival. Scores of mobile toilets and phone booths had even been installed as a kind of infrastructural support for its existence. The new pragmatic attitude of the URA and the police could only stem from the synergy that they saw in the phenomenon—the market served the needs of the foreign workers, who bought the cast-off goods at rock bottom prices, partaking of the sub-economy for the recycled goods which would otherwise go to the dump. The workers surely found the existence of the market handy—so the next best thing the authorities could do was to officially launch the market, re-name it, and make it legal, so that some measure of propriety would preside over the informal exchange of money and goods. The drawing away of the foreign workers to the open ground, which has yet to be redeveloped, also relieved to some extent, the rife congestion in Little India, which does not offer large green spaces in its hold. The whole arrangement had worked out to the satisfaction of all parties involved—the police, the vendors, the foreign workers, and the residents of the district, and at little expense to the URA.

However, today, most of this informal public space and open spaces surrounding Weld Road has been re-claimed by the government in order to construct Jalan Besar MRT, which is slated to open in 2017 as part of the Downtown line. This has resulted in the Weld Road Sunday market shrinking to an area that is smaller than ever, teetering on the brink of obsolescence. In turn, other spaces have opened up, such as the car park next to Broadway Hotel on Race Course Lane that is now used as a makeshift flea market—temporarily closed to cars on Sunday evenings (Fig. 3.22).

When I visited the Sunday Market in the early 2000s, I found myself peering over the shoulders of some excited foreign workers for a Kabaadi tournament—at the green space next to the market—a popular Indian game played in rural areas of Punjab. The tournaments were organized regularly by the same people who organized the Sunday Market—the Sunday Market Working Group, a quasi-government group set up to deal with the congestion in Little India. At other times, free Hindi movies were screened, as well as variety shows and Karaoke competitions, introducing entirely new spatial practices to the otherwise vacant plots of land. Unfortunately, this open space no longer exists today due to the MRT development. Said Dr Lee Boon Yang (former Minister for Manpower) at the launch of the Market in 2001, "The Sunday Market is a small gesture to make their (foreign workers) stay in Singapore more pleasant". At a public housing development within the district, Lee put the initiatives a little differently:

> Speaking to reporters after a block visit to Rowell Court in the Jalan Besar- Serangoon Road area, a popular hangout of Indian foreign workers, Dr Lee said he took a 'very active interest' in the large number of foreign workers in Singapore. As an interim solution, he said, weekend bazaars have been organised for the foreign workers so that they would not loiter at Housing Board blocks (The Straits Times 2001).

These new types of spatial practices have emboldened the foreign workers to improvise new uses for space in looking for new sites for new activities on their own initiatives, such as playing cricket at Farrer Park, a green space not far from Little India:

Fig. 3.22 Sunday market clockwise, from *bottom left* Construction of Jalan Besar MRT station on Weld Road, some of the knick-knacks one might find at the Sunday market, and the makeshift flea market at the car park next to Broadway Hotel, and the remnants of the Weld Road Sunday market. *Source* Author's collection

> It's more Lagaan than the World Cup, more sub-continental village cricket than a one-day international. But it is an international tri-series of sorts nonetheless. Like the heroes of the Oscar-nominated Hindi movie, these foreign workers overcome the odds - to organise their own cricket matches in a country where the game is little known. Indians, Sri Lankans and Bangladeshis get together to bat, bowl and chase the ball here on weekends.
>
> Some players come in shorts and slippers. Some are barefoot. Nobody uses protective gear. They use a rock to make holes in the ground for the three wooden stumps. Some among the spectators keep the scores, sitting on the ground, on the sidelines. But there's no mistaking their enthusiasm. Said Mr Elegan Ganesh, 28, from the Indian city of Chennai, formerly Madras: 'In India almost everybody loves cricket. Whenever there is a match in Madras, thousands watch. 'There's nothing more exciting than cheering your favourite team. In Singapore, not many people enjoy cricket. But we prefer to play cricket because it reminds us of home.'
>
> They like the venue too. 'It's so crowded on Serangoon Road, but spacious and windy here. We can still catch up with friends after the game,' he said. Some of the thousands of foreign workers, who throng Serangoon Road to chat, eat and shop on Sunday evenings walk over to the field where the cricketers play. There are usually some 150 people watching. When there are matches between 'national' teams - say, if the Indians are playing the Sri Lankans - the crowd can swell to as many as 400.

> The Indian players said they had been playing at Farrer Park for about six months. Foreign workers play cricket at other locations also, such as Kaki Bukit, Jurong, Kallang and Geylang. But they haven't always been welcome. Mr Ganesh and some of them had once played in a field in Geylang before they were asked to leave (Straits Times 2004).

However, this practice, too, has died down since the 'Little India Riots' of 2013, as police have begun clamping down on the amount of space available for migrant workers to co-opt and to congregate. Spaces in Little India are hotly contested and public space that caters to the rest and recreational needs of the migrant worker are limited; however, so long as migrant workers keep arriving in throngs, there will always be a demand for Sunday markets and informal spaces in the area. It seems like this informal sub-economy will continue to persist despite transport developments and heightened surveillance of such public spaces.

3.3.9.3 The Role of NGOs

Another type of boundedness in Little India that is often not as apparent, is the bounded solidarity that is offered by non-government organizations which fill the gaps that the government and the Ministry of Manpower have created with their strict foreign labor policy. One such organization is Transient Workers Count Too (TWC2), which extends assistance to workers like Satish Katakam who face forced repatriations, unpaid salaries and employers refusing to bear the cost of medical fees. TWC2 provides a shelter to injured migrant workers in Little India, and as part of their 'Cuff Road food program', offers free meals and serve an average of 200–350 workers a day.

On-the-ground organizations such as TWC2 create a network of relations between migrant workers in Little India, and play a significant role in re-establishing identity, belonging and ownership to the spaces within Little India. Most importantly, they give a voice of representation to these marginalized 'invisibles' in Singaporean society, especially due to the foreign workers' "ineloquence and inability to speak the language of law" due to a "systematic muffling via interpretive and replicative mechanisms of political discourse that foreclose the power of speech and representation" (Tan 2016).

3.3.9.4 Negotiated Space

The most striking deterrent to the non-typical use of space is the exceptional police presence in Little India. They operate as chief negotiators of public space, especially on Sunday afternoons, the time when most migrant workers visit the space on their days off. While the auxiliary police have always patrolled Little India, the degree of monitoring and surveillance has intensified since the riots. These displays of power and, to use a Foucauldian term, 'governmentality', have symbolized the re-negotiation of space away from those who are deemed 'undesirable' or those who are a perceived risk, namely these migrant workers (Fig. 3.23).

Fig. 3.23 Police presence in Little India on Sundays. *Source* Author's collection

> "I have a few experiences with the police. One day, I went to church, on the way back to my room, I called my parents. I talked to them half an hour. I sat down somewhere, near Rochor Canal. Then the police came, they asked (for my) IC and why you are here, why you are sitting here? I was just talking phone with my parents, what's the problem? He just asked my IC, I showed my driving license they just write down… Another time, two of my friends (and I), we bought snacks, we sat down near the HDB block. That time the guy is already hungry, they said okay lah, let's eat. We just started eating, there is one Tamil police, few Chinese and Malay all together. One Tamil police lady came over there and said, you cannot eat here, we are going to fine you. If you eat here we are going to fine you. That time I (was) very scared, that time I was not working also. This happened one year ago. Like that, we are very scared. My friends also very scared. For eating! These things happen. Sometimes, once we finish everything (work), we ask friends to come just hang out here (in) Little India, and that time they are very worried. I ask them to come out, let's just sit down somewhere, but everywhere police surrounding. So they don't want to come" (Katakam 2015).

The foreign workers, in improvising new uses of public spaces, realize that space would always have to be negotiated. Spaces in the city for marginalized groups are inexorably bounded—there is ultimately a limit to the generosity of the authorities. Whether through police presence or through barricades, these spaces are defined by those other than the foreign workers themselves. Their space in the city is also bounded temporally—there is a time and place where they could use public space freely, but bodies and minds have to be pre-occupied with work at other times. The police have conducted round-ups in the past on a daily basis of foreign workers who slept out in the open fields at night. However, the phenomenon of the Sunday enclaves in public space have also pushed the envelope of tolerance, both of Singaporeans and the state, into a form of "pragmatic benevolence" (Naidu 1999, pp. 134–143). The persistence of the Sunday enclaves and the need to deal with these creatively have, in fact, transformed public space and spatial practices in this instance.

3.3.10 Ethnicity, Exchange and Boundedness

Little India as a distinct district had developed 'organically' through history—its entity as a planned ethnic quarter had only been a recent development since the Conservation Masterplan of 1989. Changes had occurred in response to opportunities and challenges that arose and were often caused by forces from beyond the bounds of the quarter. That it had developed as an Indian quarter had to do more with circumstance than plan. So too had the various aspects of public space discussed here, as these are constantly shifting and adapting to new circumstances that arose. While some functions of public space in Little India had persisted throughout its history, new roles have also been assumed.

While public space has played a historical role in representing culture and in establishing identity, the definitions of "insiders" and "outsiders" continue to change in relation to one another, usually by degrees rather than by sharp differentiations. The public space of Little India does not show up as a monolithic entity, but as contested territories and polarized zones, depending on the beholder. The globalization of labor also has had an impact on the district, but its effects are an inversion of what is frequently regarded as its homogenizing effect. With the influx of temporary foreign workers from the Indian subcontinent, 'Singaporeanized' Indians come into contact with the spatial practices of those from more rural and less modernized parts of India, such as through the re-introduction of games played in Indian villages. They interact with—or are at least exposed to—people whose needs are more basic, and who see public space as a resource that helps to meet many of these needs, for example, of companionship and for sharing information.

As a space of exchange, we discover that Little India offers "front" and "back" public spaces, if one can read the indicative signs and signifiers legibly. These environments are not always defined as demarcated urban configurations, but they are often regarded as an overlay of overlapping spatial fields. The absence of women and children from the public spaces of the area compels one to wonder why this is so, and what could possibly be done to bring the absentees from public spaces to the foreground. These absences also highlight the lack of urban public space that would draw out certain segments of the population. In spite of this, spectacles such as the annual Thaipusam festivities draw women and children to public space, as these are celebrated by many Hindus, with roles for women and children. As a lived space, the district is a divided locality, inhabited by an imagined community, which nests subversive spaces and heterotopic spheres within its boundaries.

The new spatial practice of dense congregation, involving the logic of a seemingly chaotic crowd, engendered new forms of negotiation with the hegemonic forces of the state, forcing creative and sometimes seemingly oppressive resolutions of the problem. Temporally bounded spaces are colonized by the weak, who are silent but numerous and persistent, who come up against the 'vocal' protestor, for example the residents of the quarter who are intimidated by the congestion, but otherwise absent from public space. New devices for containing the crowd in public space have emerged, ranging from creative dispersal, negotiation, and the mediation of differences, to increased forms of regulations and tightened security. Negotiated thresholds of publicness often transform spaces into sites of tension, or intimate friction. Public space in Little India embodies structures of boundedness embedded in its urban fabric, not least in the tangible forms of barricades and fences. New forms of improvised spatial practices have emerged, not just in the tactics of marginalized groups, but also in engendering the new spatial practice of pragmatic benevolence by those in power.

The 2013 'Little India Riot' has left an indelible impact on public space in Little India. The newfound visibility of the foreign workers in public space in no small part contributed to heightening the social problems associated with importing labor into Singapore. The riot reminded the public that the foreign workers are by no means a statistical and abstract figure, but are real bodies occupying real space on a small island. The case of Little India had also demonstrated a coming-of-age of a dormant civil society, in the form of Singaporeans who have stood up and spoken for marginalized groups like the foreign workers, and non-governmental organizations who have provided some form of support and relief for these groups such as free meals, the provision of forms of recreation and entertainment, and making available affordable goods. It also reveals the careful balance that the state tries to strike between its planning priorities and serving its stakeholders, as it seeks to preserve cultural heritage in order for Little India to remain a healthy neighborhood community, a viable tourist destination, and a manageable public space for its weekend users—in that order. The whole process is dynamic and changing, reflecting shifting power roles and the practice of negotiating through practices within public space.

3.4 Public Housing Spaces: Spatial Practices in Transition

> The heartland implies by its name that it is the core of Singapore. Geographically the heartland is spread out all over Singapore, but in its representations it is a space that denotes national and local identity beyond coordinates. It is an exclusionary space…it is a myth of origins, of a belief that history and landscape that are exclusively Singaporean exist. They are nationalist narratives, taking on "the naiveté of nativity: the pure, true national story that is pure and true because it is native" (Yeo 2004, p. 27).

Although the form of public housing adopted by Singapore from the 1950s onwards was not unique at that time, the way spatial practices have developed within the new environment, the so-called Singaporean 'heartland', took on a unique shape —a condition the excerpt above terms as 'native'. How spatial practices had persisted in the new spaces, how some of these practices had differed from the past and how the design of space had influenced spatial practice, all developed conterminously with the social culture of the Singaporean brand of public housing. A review of the history of Singapore's public housing and the development of public spaces within this framework would reveal a history of experimentation with new spatial typologies that form an armature of potential social spaces. As public housing spaces had been cumulatively built, they exist in contemporary Singapore simultaneously, irrespective of their vintage, and as a collective present, a rich spatial field for investigating the roles of public space in contemporary Singaporean society.

While the spaces of public housing in the early days were functional in nature and served to provide light and air rather than act as social spaces, latter spaces in neighborhood clusters and in precincts were devised with some kind of social agenda (such as 'building communities'). The town centers have also served as social centers of new towns, with their concentration of amenities and as transport hubs. The hierarchical progression of spaces in the new towns, their even distribution and their increasingly bounded nature might be read as concrete manifestations of Confucian ethics and the paternalistic nature of government in Singapore—where a predominantly Chinese sense of authority and control still persist. While places of commerce as well as the traditional urban fabric were shaped through evolving spatial practices in the city, the spaces offered by public housing were created quite directly through government policies and new town planning agendas. These spaces nevertheless mark the datum of experience of public spaces for many Singaporeans, as eight out of ten in the population find home in a public housing flat.[58]

3.4.1 *Framing Public Housing*

The discussion of public housing spaces will be first approached within its macro-context and historical development, and then within specific instances of

[58]From the Department of Statistics Singapore website on 'Households and Housing Data'.

forms, space and spatial practice. The study will begin by tracing interrelationships between the political agenda and choices made by the state, and the ensuing design choices in public housing in the Singaporean context. In the case of new town spaces, urban design politics is dominated by the hegemony of planned spaces. However, at the level of spatial practice, the design politics of the macro-level is not always translated to prescribed use or desired social outcomes. Public space in public housing makes manifest the interpretive nature of spatial practice within the designed environment. This chapter examines in particular the roles of public space in relation to the forms of spatial practice in these instances:

1. As space where social practices have persisted, with these practices sometimes taking on new shapes, but often easily adapted to the new spaces in public housing.
2. As space where new social practices have occurred, due to the nature or distribution of these spaces; where new forms of spatial practices have been created through the interaction of people and the built environment, as well as instances where new spaces and new practices have sprung up.
3. While the forms of control of public spaces tend to reinforce the status quo, subtle forms of defiance have occurred, and cumulatively provide some form of resistance to an idealized picture of the public realm. However, there are also instances of less "polite" versions of defiance in public space—such as the phenomenon of "killer litter"—aerial projectiles constituting everyday objects thrown from a high-rise flat into the public spaces below.

The spaces of public housing shaped from both above, through planning and design, and below, through spatial practices, capture constructions of Singapore public space in transition. To understand the context of public space in public housing, the history of Singapore public housing will be outlined in brief.

3.4.2 Singapore Public Housing—A Brief History

The planning and urban design choices made during each decade of public housing —from the colonial attempts at a public housing prototype, through the inception of a massive public housing program from the 1960s to the present[59]—parallel the state's attempt to socialize the population towards desired social and political goals. The role of public space had not always been of prime concern in the history of public housing, but its importance had emerged in recent years in tandem with the emphasis on "building communities".[60] This trajectory in the development of

[59]Public housing in Singapore is implemented as 5-year building programs, and as such, a decade is a good time frame for review as it covers two building programs, and goals and orientations become clearer as they are sustained within the consecutive programs.

[60]This phrase is reflected in the current mission statement of the HDB.

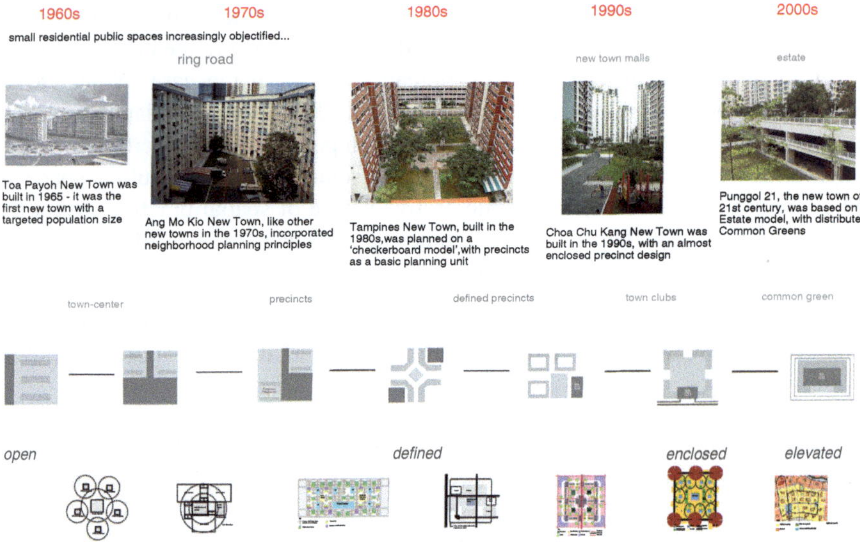

Fig. 3.24 Development of new town spatial typologies

public space is itself illustrative of a change in the thinking of the polity with regards to space as a social resource.

The public housing of today is far from the early days of cramped, low-cost housing, when the call of the day had been to "break the backbone of the housing shortage" (Liu 1992).[61] To date, the Housing and Development Board (HDB) has built twenty-three new towns with a total housing stock of more than 965,000 flats.[62] Public housing in Singapore has none of the stigma associated with it as it does in North America, and is a widely accepted as the form of housing for the lifetime of a lower-to-middle class Singaporean. Currently, 80 % of the population lives in public housing. Of the 5.5 million population expected by 2040, 75 % of the population are projected to dwell in public housing.

However, with growing affluence, higher expectations are made on the quality of the housing environment. The spaces of public housing have shifted in function from pragmatic spacing between buildings and as leftover spaces, to become the focus of planning for "cohesive communities"[63] and also as tangible images of the improved quality of life (Fig. 3.24). The transformations of spaces in the design of

[61]Liu had definitively stated that the housing shortage was solved by 1978 and the phrase was not used again after that year.

[62]These statistics are derived from the website "HDB Towns, Your Home", accessible at http://www.hdb.gov.sg/cs/infoweb/about-us/history/hdb-towns-your-home. In 2000, HDB welcomed its 800,000th homeowner in a symbolic key hand-over ceremony celebrating the organization's 40th anniversary.

[63]This is the term used in the current corporate vision of the HDB.

new towns is most impactful at the level of small residential public spaces—as these are the most immediate public spaces for the everyday life of many flat dwellers.

The poem, "A Brief History of Toa Payoh" by a local poet Koh Buck Song strikes a chord with many Singaporeans, as it embodies the swiftness of change through the history of Toa Payoh New Town. The New Town was a brave new world of its time and a promise of progress in comparison to the swampy, unsanitary homes before:

> ...was once a brave new world
> of giant pigeonholes sky-looking,
> rigid under morning sun,
> a pioneer in its time

But the generation after those pioneering flat dwellers moved to even newer, improved towns, endorsed by the nation-state as a sign of progress. The older towns were overtaken by newer, 'better' towns just as the children of pioneer flat dwellers had superseded them in wealth and educational status:

> the pride and self-sufficiency
> of early settlers
> eclipsed:
> town centre, bus termini, the first SEAP Games,
> the emporium's sacks of fragrant rice
> children's playgrounds, the garden's lake and tower,
> the Queen's lookout
> excite no more
> and the children of Toa Payoh
> are the mothers and fathers
> of Woodlands, Pasir Ris,
> distant orbits of the new satellites (Koh 1992)

The decades for which we trace the development of public housing are thematically articulated here to reflect the political agenda embedded within the design and planning of each decade.

3.4.2.1 Shelter (1952–60)

> When HDB was formed in 1960 to solve the massive problem of substandard and insufficient housing, it signaled the beginning of a lifestyle change for Singaporeans, from kampong living to high-rise and high-density living. Going upwards was a necessary move as our population was growing rapidly in the post-war years.[64]

---

[64]Excerpt from the speech by the Minister for National Development, Mah Bow Tan, at the groundbreaking ceremony for the first 40-storey apartment block at Toa Payoh in Jun 2001.

Although the decision to adopt the high-rise, high-density housing to alleviate the grave housing problems of the 1950s and 60s remain largely unquestioned,[65] it is worthwhile to reflect on the context of the decision to do so. Singapore had remained a British colony up to 1959. In the early 50s, the colonial government, through the Singapore Improvement Trust (SIT)[66] had just started its program of building estates like Tiong Bahru[67] and Queenstown. Many of the ideas about housing being implemented in Europe in general and Britain in particular, found suitable testing ground in Queenstown, where the high-rise model was first adopted,[68] initiated by the SIT in 1952 and completed by the HDB, the local housing authority in the 1970s.[69] The estate had a population of 150,000,[70] formed by the amalgamation of smaller estates, with a town center added later, to become the first satellite town. The urban fabric of Queenstown broke away from the street-based city blocks of the shophouse-type, which was the predominant urban form of the city center at that time, to primarily orientate the estate in the north-south direction with green space in between blocks to fulfill lighting and ventilation requirements (Fig. 3.25).

Economists have argued that low-profiled dwellings produced better and greater use of space in contrast to HDB's high-rise, which wasted space between buildings (Pugh 1984, p. 44). This begged the question of whether Singapore had accepted such a model of new town planning from the West at a time when the urgency of the housing situation overtook the need to examine other implications besides the technological feasibility offered through the high-rise, high-density model; and at a time when high-rise livability was being questioned in the western countries.

The policy to resettle families away from the central area, which was to be developed for commercial purposes, and the systematic acquisition of parcels of land for new housing favored the building of high-rise flats. The high-rise option also provided a viable symbol of progress and improvement with the new self-government (Liu 1982 cited in Jayakumar 1982, p. 134).[71] Since its inception in 1960, the HDB had turned to models like the British new towns, but with the

---

[65]This is the case, except in academic circles.

[66]The Singapore Improvement Trust was the municipal body set up by the British colonial government in 1927 to "provide for the Improvement of the Town and Island of Singapore" (Fraser 1948).

[67]Tiong Bahru Estate was built by the Singapore Improvement Trust (SIT) between 1936 and 1941, with later additions in 1948–1954.

[68]The high-rises were blocks that rose up to 10 storeys.

[69]The HDB was set up in 1960, after Singapore's self-rule, to oversee the task of housing the growing population.

[70]Net residential density was high and ranged from 800 to 1000 pp/ha (Tan et al. 1985, pp. 56–112).

[71]This was dramatically underscored when the then Prime Minister, Lee Kuan Yew visited the site of the fire of May 1961 at the squatter settlement of Bukit Ho Swee and promised that blocks of flats would rise form the devastated area with the first units ready for occupation within nine months. Exactly after nine months, five blocks of flats had been built to re-house 800 families.

Fig. 3.25 HDB blocks at Bukit Merah. Examples of high-rise HBD estates located within the Queenstown planning area. *Source* Author's collection

need to house 150,000–300,000 people in self-contained new towns, it was felt that "(t)here were no useful models to follow" (Liu 1992).[72] The ground was thus primed for the invention of urban models for new towns "customized" to Singapore's political and social context.

3.4.2.2 Consolidation (1961–70)

Public housing was seen as a means to allow for the hollowing out of the central areas for commercial development, with the resulting need to cater for the massive resettlement of the population. Toa Payoh New Town, initiated in 1965, was the first new town planned with a target population of 180,000 and was in fact HDB's flagship project for public housing. The rampant housing shortage at the time called for functional design and quick implementation of the new town. The model of housing was very much based on the Congrès Internationale d'Architecture Moderne (CIAM) model of functional mass housing, with its integrated living,

[72]A case in point would be to compare the density of Milton Keynes at 28.1 persons per ha with that of Toa Payoh New Town—which was the first new town with a target population, at a density of 1146 persons per ha.

transport, work and recreational components.[73] The ring-road system was used, and the town had a center, which included commercial, institutional and transport facilities. Spaces between buildings were simply turfed over and served pragmatic functions of allowing light and ventilation, the occasional playground being provided where space allowed.

3.4.2.3 Social Disintegration and Re-integration (1971–80)

> Singapore's multi-story housing of the 1960 s was purposely anti-communal. Although there were many ritualistic references to "building new communities" in early HDB literature, the far more urgent problem was the breaking up of an existing communalism seen to be identical with slums, kampungs, and squatter camps. The HDB's policy of purposely mixing "races" in housing estates and even individual buildings have been well documented and well-discussed. But this policy of mixing extended also to pre-existing communities of the same "race"...The HDB system would in effect atomize preexisting sub-racial communities..." (Clancey 2004, p. 45).

Although public housing presented a far superior living environment than what it replaced, there was an undercurrent of dissatisfaction with the failure of the HDB to protect a sense of community, neighborliness and identity, and with the breaking up of the traditional extended family structure (Yeh et al. 1972). On a more insidious level, this could be read as an attempt to atomize already existing strong communities of the same race, whom in Singapore's history had initiated strong communal lobbies, protests, and even inter-racial conflicts. However, the government could not be seen prima facie to be the spoiler of communities. These were some of the issues that the public housing of the 1970s tried to deal with.

Throughout the 1970s, as the public housing programs were consolidated and housing provisions brought up to an adequate level, the political will turned to the planning of new towns with neighborhoods of convenience as a functional model of distribution of infrastructure and facilities. The new towns were also planned to provide some means of social integration for a society dislocated from its vernacular and ethnic origins—addressing the perceived need for some form of reintegration and neighborliness to develop. Neighborhoods were conceived as self-contained "communities" defined as "total living environments" of about 6000 dwelling units,[74] sufficient to support a primary school, shopping, and community activity nodes within a walking distance of four hundred meters.[75]

The early neighborhoods were planned as centers of convenience, not centers of community building, and were executed at a very pragmatic and functional level

[73]The proposal of the CIAM at its 4th Congress in 1933 produced "The Athens Charter", a document which remained controversial in its commitment to the planning of functional cities, high-rise housing, zoning and green belts.

[74]This is detailed by Tony Tan, Chief Architect of the HDB in his public lecture at the Department of Architecture in September 1999.

[75]The area of each neighborhood was about 40 ha.

(Liu 1975). To ensure full usage of neighborhood centers, these were spaced apart at 900–1200 m, "so that a resident does not have any choice between centers but can reach them on foot" (Liu 1975, p. 152). For residents living beyond 230 m from the neighborhood center, neighborhood sub-centers consisting of five to seven shops and an eating-house were provided to serve the locale.

Ang Mo Kio New Town was constructed in 1973 based on a prototype new town model. The hierarchical model of a town center serving neighborhoods equidistant from the center, and in turn, served by neighborhood centers and sub-centers, was adopted on a large scale. The blocks were typically arranged in a north-south orientation, which was one of the important layout criteria for flats with the emphasis on functional planning. The buildings do not form enclosures for the spaces in between, which were visually very open. Since the spaces were laid out as functional spacing and as carparks, there was little hierarchy in the scale and types of spaces. The spatial arrangement was generally loose due to the space being defined mostly on its longitudinal sides via the slab buildings. The presence of void decks (the space below the flats, which were raised by piloti) also made the space at ground level even less defined.

Although the new town was generally green and pleasant with neighborhood parks and conveniences distributed hierarchically within the new town from the town center to the neighborhood centers, there was a perceived lack of a social dimension to the space. The sameness of spaces created through the standardized designs of new towns became apparent and the designs were subjected to growing criticisms. A new model of planning had to be derived to address this problem.

3.4.2.4 Building Communities (1981–90)

The government realized in the late 1970s and early 80s that the design of public housing and its spaces would be instrumental in its new agenda of social engineering for a population largely housed in HDB flats. The abstract notions of 'building communities' and the optimization of land took concrete form through the implementation of the precinct concept, which created opportunities to foster a sense of identity and belonging through the inward focusing and compact precincts, which form the interlocking lattice for the new town.

In parallel to the development of the precinct, the government saw the building of new towns as the perfect opportunity to encourage an even greater extent of interracial mixing (or of greater racial distribution). During the first massive relocation program, it was observed that racial groups were congregating in particular housing estates, such as the Chinese in Ang Mo Kio and the Malays in Tampines. Explicit controls in the form of racial quotas in each new town based on the current demographic quota of the nation—enacted even at the level of each block of flats—were introduced in 1989 through the Ethnic Integration Policy. In this way, there was a deliberate mixing of classes at every level of the new town through the mix of flat types within neighborhoods, precincts and each block of flats.

As a social focus, the precinct space contained shared amenities and increased the chances for propinquity among racially-mixed residents who come from different sizes of flats (and presumably different economic status), with its sheltered spaces and soft landscaping, and subsequently, multi-storey covered car parks. The creation of identities for precincts and thus for the new town as a whole required planners to pay more heed to natural features and topography, as well as to create customized designs for precincts.

The New Town Structural Model made its first appearance in the late 1970 s and formed the basis of the "checkerboard" model of town planning, characterized by the use of the precinct as a basic unit of planning (Tan et al. 1985). It repeated itself in clusters of 4 hectares or sometimes half the size, serving 400–800 families housed in four to eight blocks of flats. Each precinct had a precinct center, which might include small games courts, children's playgrounds or landscaped gardens. The precincts acted as structural elements in the new town plan, and could be interlocked and combined to create rhythms and to relate to different categories of roads. They could also circumscribe facilities within courtyards and open spaces (ibid.). Six to eight precincts would share a neighborhood center, which was usually within a four hundred meter radius or five minute walking distance from the precinct. Subsequently, the emphasis of public housing in the 80s was to create new towns with a stronger image, a stronger sense of community and identity.

The idea behind the creation of the precinct was to try to encourage meaningful social interaction among residents (now deemed as lacking in neighborliness) through the shared use of this focal point of activity at a scale that residents could recognize and understand. The precinct space and facilities were planned with the "expectation that residents using them will come into social contact with each other, and in so doing, develop a sense of belonging to their individual precincts. Unlike the other physical planning elements, the precinct is a spatial engineering strategy to effect the achievement of social goals" (Teo 1996, pp. 279–294).

The first precinct design was implemented at Tampines New Town in the early 1980s and consisted of a children's playground, games-courts, landscaped areas and local shops. Here, to encourage residents' perception of the precinct space as a locale, "clear definition of the precinct boundary has been attempted…and blocks orientate towards a central open space" (Ooi and Tan 1992, p. 80). This was in contrast to earlier arrangements where playgrounds tended to be located near roads or at interstitial spaces to avoid noise disturbance to residents.

On a sociological scale, the precinct seemed also to be the response of planners to the findings that the neighborhood of familiarity operated on a much smaller scale than the neighborhood of convenience used in the planning of the new towns (Tan et al. 1985). It had also been found that "the frequency of use and the significance of the range of public spaces in the daily lives of residents appear to be inversely related to the order of the planning hierarchy itself" (Ooi and Tan 1992, p. 80). In other words, the spatial hierarchy of starting from the center to the periphery in planning terms was inversed from the point of view of the residents, whose daily lives might significantly involve only the peripheral spaces in the

decks were not built to advocate certain types of uses; this open definition allowed people room to negotiate and define their own space. Precinct pavilions, on the other hand, were designed as spaces with a specific function to serve the community's needs. However, with this utility comes a loss of versatility that void decks provided. The flexibility for residents to define their own space has been somewhat clipped as town councils hold jurisdiction over the types of uses permitted in the precinct pavilions. Furthermore, since the use of precinct pavilions are prescribed, the opportunities for interaction at these pavilions are therefore limited.

Precinct pavilions bring about a distancing of events that usually take place in 'one's own void deck', as void deck conversations do not automatically migrate to the pavilion. There is inherent value in the passive negotiations that take place over the use of the void deck, as neighbors had to be aware of each other's activities in order to be sensitive to each other's needs. The effectiveness of the void deck in creating social cohesion was its prime location, capitalizing on the natural flows of people leaving and returning to their homes. Precinct pavilions do not achieve this, as due to their inconspicuous location, they are not a part of people's everyday routines. As such, they are often underutilized spaces. Void decks allowed for the shared negative externality of noise pollution created by funerals or weddings to being shared amongst those living in the neighborhood, with each flat of residents taking on the role of ceding its void deck space for a short amount of time as a mark of respect to a neighbor living in the same building. Now, those who live by the pavilion experience more noise pollution due to the reduced corridors of activity within bocks. Moreover, this creates a sense of alienation between those whom the ceremony is not visible to, and those involved in the wedding or the funeral. However, this is perhaps mitigated by the fact that most homebuyers in new estates are part of a younger generation of HDB buyers who are growing in affluence and may be ditching traditional norms on how the void deck is customarily used—many now choose to hold wakes in funeral parlors and weddings at hotels.

The ability of the void deck to serve a multitude of functions within the confines of each housing estate is what made its function so welcomed and its potential so captivating. It was an antithesis to the hyper-planned, cookie-cutter apartment configurations that sat above it. Void decks are undefined—but managed—spaces waiting to be used, as opposed to its name that suggests it remains mostly void. People have the autonomy to transform the space and did so whenever they utilized it. Today, the precinct pavilion is less utilized since it is out of the way, and thus, there is not as much exploration outside its intended use. It is a space waiting to be defined by its users, but it is currently the true 'void' deck.

3.4.3 Urban Design Politics Versus Spatial Practice

The history of public housing recounted in terms of urban design politics shows firstly, a pragmatic layout of spaces, initially given hierarchy as functional types, but increasingly partitioned to create notions of neighborhoods; then smaller communities or precincts, then estates or common greens, with an accompanying re-allocation of shared amenities to each of these individualized communities than on a shared neighborhood basis; and subsequently the packaging of these spaces into semi-private domains, suggesting the notion of private properties. Such a reading suggests that design was increasingly manipulated towards political ends, which are quite explicit, but these have not always created successful spaces or environments that have met planned intentions.

However, it is only at the level of spatial practice that one can begin to understand the meaning of these spaces in the everyday life of public housing dwellers, and whether they form a meaningful notion of public space in public housing. As such, I found it necessary to move in from the macro-scale of planning and urban design to the scale of spatial practice—how people inhabit and move through public space, their interactions therein, how these become a part of their everyday lives, and if these practices lead to transformations of space, or are transformed by it:

> In representations of HDB living by different writers and filmmakers a common feature recurs: the magnifying of everyday movements, sights, and smells in an estate. In the films a favorite frame consists of blocks of HDB flats and their corridors lit by fluorescent tube-lights in the night. Also popular are scenes of residents going about their everyday business, sitting and chatting in the void decks, or taking the lift and walking down the corridor to their flat. There is a grasping of the local and the Singaporean communal in these scenes… (Yeo 2004, p. 25)

To grasp the '*Singaporean communal*', I have framed instances of spatial practices within the three frames suggested at the beginning of the chapter.

3.4.4 Persistent Spatial Practices

The design of HDB apartment blocks, elevated off the ground on pilotis to create a sense of flowing space (at least at ground level) and to alleviate the feeling of oppressive density, leaves a blank space interrupted by columns and piers on the ground floor. This space, simply called the "void deck", plays an important role in structuring the spatial practices of the everyday life of residents. For many, it is the point of arrival and departure from their homes—a part of the system of void decks, lift lobbies, corridors and landings in the transition from home to the outside world. It is the point of sending schoolchildren off, waiting for those on the way home, and of informal social encounters, as embodied in selected excerpts of the poem, "Void Deck" by Alfian Saat (1998), a Singaporean poet:

> Where the neighborhood wives,
> After a morning at the wet market,
> Sit facing the breeze
> To trade snatches of gossip…

Naturally, the inhabitants of void decks include children who engage in wanton play:

And children orbit around them
Laugh without diction –
Their games of tag a reassurance
That there has been no hothousing
Of who is unclean, unwashed,
Untouchable…

And the older inhabitants of the void decks, who do this:

…sit like sages
To deploy chess pieces with ancient strategies.
In a corner, a caged bird bursts
With the song of its master's pride…

The system of spatial transitions of corridors and void decks appealed to the cultural habit of the Singaporean dweller of public housing in extending personal and cultural space outside of the threshold of the home. The extensions in question take the form of the physical extension of personal artifacts, like footwear, plants, religious paraphernalia, including altars, incense burners and graphic representations of the Hindu Mandala drawn in chalk outside of the home; furniture, including benches, seats, umbrellas, bicycles and so forth. This phenomenon is eloquently illustrated, again in an excerpt by Alfian Saat, termed "Corridor" (2004):

Upon exiting the house, the first space I encounter is the corridor. I have two other neighbours sharing this corridor, namely a Chinese one that inhabits a remote unit at the end, and an Indian one adjacent to my flat. The act of possessing this public space involves various strategies, which on the outset appears to be merely aesthetic, or even economic, if one perceives such claims to be an attempt at the extension of floor-units.

The Chinese family has marked ownership by the literal purchase of a segment of the corridor; this they have converted into a transitional space which one could term a patio or veranda. The outermost limit of this space is a gate that fills in the outline of the corridor; through this gate one spies the occupants' shoes, umbrellas, etc. These objects are ostracized from the interior of the house (the umbrella is wet, the shoes' soles are unclean) yet also protected from the occasional pilferer, by virtue of their position as being simultaneously inside and outside. Both privacy and security seem to be motivations for imposing this barrier, but the elevation of status plays a role as well…

As for my family and the Indians next door, dominion is expressed through the arrangement of potted plants as well as shoe racks. They act in lieu of actual fences, although the prohibitive dimension of such boundaries is mitigated by the notion that they represent a spillage of the actual items (furniture, household botany) populating the home, a riotous excess of the interior. Yet there is counterpoint: like compass needles, the footwear (and the doormat) are oriented towards the direction of the door, a sign that attempts to reverse the traffic of overflow. In other words, to maintain an ecology where the luxuriance of objects does not displace the public character of the corridor; anyone who traverses a HDB corridor will thus not have his or her conscience tainted by notions of trespassing. The space of the corridor is apprehended, yet also free: a continuous footpath running across other people's gardens.

Cultural extensions of space also include the 'exteriorizing' of family events such as weddings, funerals and birthdays, bringing these to public space. It is already an HDB tradition in its history of just over forty years, that Malay weddings were held in the void decks of flats. The extension of cultural practice to public space had, in my observation, been the strongest in the Malay ethnic group.

Despite the efforts by the government throughout the history of Singapore public housing to effectively disperse the ethnic groups, the Malays had maintained a sense of community and ethnic identity in several ways. Through the buying of re-sale flats, the Malays had generally been able to acquire as an ethnic preference, most of the lower level flats, which the Chinese majority of flat dwellers, when given a choice, did not prefer. Living on the lower floors gave the Malays greater accessibility to the ground floor space and to have other Malay neighbors. Furthermore, the spatial relocation of mosques into new towns played a role in shaping religious identity and kinship networks among Muslims living in HDBs. Men living in the same neighborhood usually visit the same mosque and consistently interact with each other during fixed prayer times. A representative "headman" type person is also informally selected within each block to resolve any matters involving the Malays. The void deck acts—not by drawing a simple vernacular parallel but more alluding to a persistent spatial practice in new spaces—like a *kampong* (village) compound and the corridors as elongated common *serambi* (verandahs). The Malay wedding in the void deck is very much a communal affair, with neighbors and friends helping to cook, wash and in hosting the event:

> Guests are invited to partake of a meal on Sunday. This is usually held in the void deck of a housing board flat so as to accommodate the large number of guests invited. Besides cutting down on costs, holding the feast in the void deck also enables the guests to view the bridal chamber and the bersanding (sitting in state) ceremony often held in the pengantin's (bride/groom) home. The wedding preparation is often based on the gotong-royong (co-operation) among friends and relatives, for which the Malays are most well known for.[78]

The void deck is quite often used for Chinese funerals, the wake for which lasts from between three to seven days, and the proceedings range from visits by friends and relatives to pay their last respects to elaborate Taoist rituals involving musical accompaniment and the burning of paper offerings, for example, in the form of a large papier maché model of a condominium block.[79] Temporary tents, mobile toilets, sound systems, lighting, seating and tables, as well as an elaborate backdrop for the coffin form the paraphernalia of the Chinese funeral, filling up the void deck. The whole set-up of the funeral temporally transforms the void deck into an extension of the Chinese family home, and strangers who pass through the space are treated with mild suspicion—the public space is temporarily transposed within the private realm. The 'publicizing' of private grief alerts the flat dwelling community

---

[78]From "The Malay Wedding"—http://www.zawaj.com/weddingways/malay.html.

[79]Burnt paper offerings in the form of material properties and "Hell" currency are rooted in Chinese beliefs in which earthly provisions from the pious could help make the after-life in Hades much more tolerable.

Fig. 3.28 Funeral at void deck. The void deck space being used to host a wake for a deceased Chinese resident. *Source* Author's collection

of a death in the family, performing the role of communication and at the same time re-enacting shared cultural identity (Fig. 3.28).

The enactment of weddings and funerals in the public space of the void decks not only causes neighbors to lose the use of the void deck temporarily, but also imposes on the private realms of other flat dwellers through the presence of crowds, the pollution of the air through burnt offerings and cooking in public, and least of all, the noise levels these events generate. However, fellow flat dwellers have demonstrated extraordinary tolerance, honed through years of dwelling in close proximity to neighbors whom one hardly knows. The tolerance of the extension of cultural practices in public space, though passive, is itself a form of spatial practice.

The tolerance of these spatial practices, through the town councils and the HDB, is itself institutionalized in its Town Council rules and regulations, in which sanctioned social functions in "void decks, link houses and amphitheatres" and include "weddings, birthdays, Hari-Raya (Malay New Year) get-togethers, religious rites, funeral rites, Terawih (breaking of the Fast for the Muslim community), block parties, etc."[80] Activities sanctioned for open spaces include Seventh Moon

[80]The Town Councils impose a nominal fee of between S$30 and S$60 for the use of electricity, rent of the space and water charges. The use of open spaces is also charged. All these activities require a permit to be obtained prior to the staging of the activity.

Fig. 3.29 Persistent practices

Prayers, Puppet Shows, *wayangs* (staged performances) and variety shows. Through the imposition of rules of behavior, charges and temporal boundaries to the events, the authorities maintain a form of control to when, where and how these take place.

Persistent spatial practices (Fig. 3.29) as outlined above continue to thrive in the system of transition spaces from the home to the public realm, with fellow flat dwellers and the government regarding these practices as benign, to be tolerated and even contributing to a sense of community. These practices may no longer take place in the vernacular space of the past but are transposed to spaces of new configurations in public housing. However, the forms of the practices persist, sometimes within a diluted but still intact social structure, and have similar meaning and content as the original.

Some of these practices persist out of sheer cultural habit, but these too are beginning to change with a new generation of flat dwellers, who are born into the HDB environment without any prior memory of deep-rooted practices and have grown up in a very different social and cultural milieu. An example is the Chinese wedding: in the early days of HDB living, young couples held wedding dinners in void decks and the surrounding spaces, but this practice has died out. Chinese wedding banquets are now held in hotels and restaurants instead, as the Chinese community has become much more affluent, and the wedding dinner is a signpost of prestige and status.

3.4.5 New Spatial Practices

The new spatial configurations of the new towns have resulted in entirely new forms of spatial practices within the new spaces, or at least led to changes and adaptations of practices to these new environments such that they are of a different form. However, this does not mean that the social meaning of these practices have changed, although entirely new practices may occur in the new configurations of spaces, how people move through them, interact, and inhabit them.

The spatial arrangement of housing alongside current policies aimed towards racial co-mingling has led to new practices that would otherwise—and in another context—be unthinkable. One such example is the *DeepaRaya* celebrations in Singapore, which take place when the Hindu *Deepavali* holiday and Muslim *Hari Raya* (*Eid al-Fitr*) holiday occasionally coincide. The shared living spaces and increased 'religious encounters' have set a necessary condition that allows for these different religious festivals to co-exist, 'producing novel religious thinking and practices' in Singapore (Sinha 2003, pp. 459–494).

A simple but significant spatial configuration in this instance is the movement system of the flat dweller in transition from their flat to the spaces outside their home. We have seen that the corridors and void decks espouse the persistent spatial practices transferred from a spatial vernacular. In the large scheme of the movement system leading from the apartment blocks to the neighborhood center and town center, a pattern changing from linear hierarchically defined spaces to a more distributed system of spaces is experienced. The following graphic mapped for Ang Mo Kio New Town illustrates this (Fig. 3.30).

One encounters more spatial choices as one moves further away from the flat. As many of the new towns built in the 1970s and 80s followed a hierarchical system of containment of spaces, one moves outwards into expanding spatial systems from the precinct towards the town center. A mapping of incidental routes for different types of people might look something like this (Fig. 3.31).

An interpretive reading of this simulated graph of different routes taken during the day by different subjects show that the places nearer home are where people most likely to run into each other. Such places, including void decks, paths, neighborhood coffeeshops, the Food Center and Market, form channels of activities, interspersed with nodes, they constitute a spatial system in which recurring spatial practices that are re-enacted on an everyday basis become embedded. Coffeeshops within the neighborhood and town centers often play the role of new village squares or

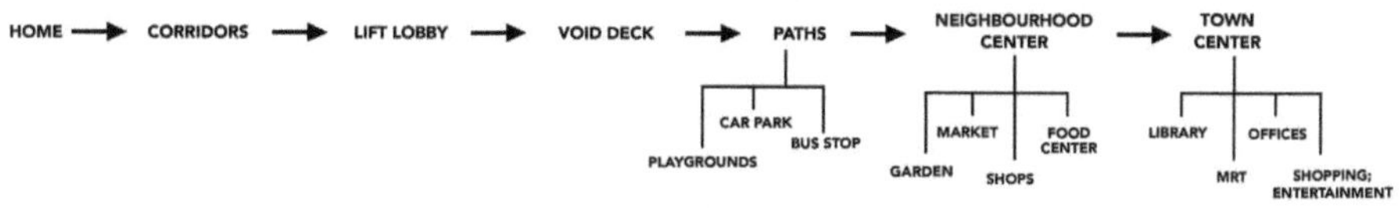

Fig. 3.30 Movement systems diagram. *Source* Author's collection

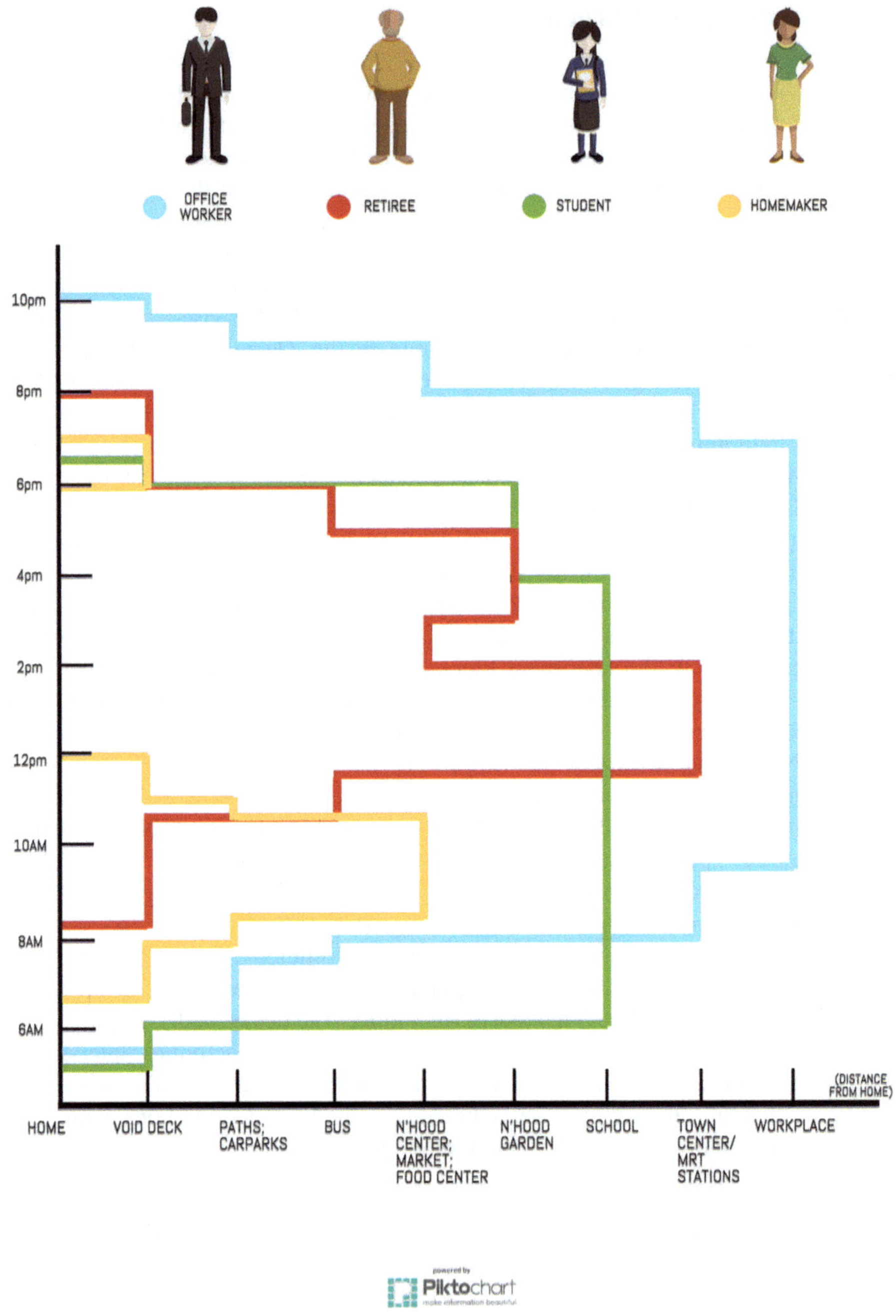

Fig. 3.31 Incidental routes. *Source* Author's collection

Fig. 3.32 New places and spatial practices

commons, with residents often meeting neighbors and friends from within the estate or neighborhood, exchanging greetings, news, or even watching important sports events together over the coffeeshop TV set. During political campaigns, coffeeshops are important campaign stops for the politicians who are keen to win the "heartland" vote. Recent social phenomena[81] such as the influx of Chinese prostitutes to the Singaporean "heartland" are ubiquitously played out at the neighborhood coffeeshops, much to the consternation of more conservative "heartlanders".

Hawker centers and markets are also places where one is most likely to meet with neighbors: groups of women are often seen chatting at the market as most of them have a routine that are likely to coincide with those of their neighbors. It is a Sunday morning breakfast routine at the coffeeshop in Tiong Bahru, where old men would hang up their songbirds on their perches and spend the morning listening to their melodic songs. A prominent void deck space near the new town library in Toa Payoh is the favorite spot for a group of regular, older, male chess players who nonchalantly flop themselves around the space of the void deck to indulge in their pastime (Fig. 3.32).

The configuration of new town centers, often tending to converge routes along the shop fronts, create intensely used paths narrowed even further by the spilling of goods along the arcaded fronts of the shops, such that an intermediate zone between the shop and the external spaces are created. The channels formed through the orchestrated routes in the town center again increases the chance encounters of neighbors and friends in these spaces, enriching the prospects of these spaces as social spaces. The spatial practice of shopping in the town center is an adaptation from that imagined by

[81]Reports of prostitutes from China soliciting in 'heartland' coffeeshops were rampant in 2004.

Fig. 3.33 A lady walking through a Mama Shop. *Source* Author's collection

planners and architects. At the level of spatial practice, shop fronts are often blurred zones of public and semi-private spaces, evoking again a cultural habit of claiming public space for private use. Plastic canopies extend beyond the neat rows of terracotta-tiled canopies provided by the HDB. Mobile carts and kiosks animate the space, and in some places such as the Kreta Ayer HDB development, vendors occasionally lay plastic ponchos on the ground to display their wares.

Town Centers and the areas around Mass Rapid Transit (MRT) Stations are the new hangouts for groups of young people, who congregate to spend time chatting or, in the case of the space around the Bishan MRT station on Saturday afternoons, meeting to exchange coveted trading cards. Instead of coffeeshops, teenagers would hang out at the new town Starbucks or meet around the new town shopping mall, where the occasional mini-amphitheater or seating ledges are provided for waiting.

Older women too, find their own niches in the new towns. Early mornings at an Ang Mo Kio precinct, a group of feisty older women meet to do pre-dawn tai chi, finishing in time for their daily market shopping. Groups of Indian women meet round a landscaped garden in a Bishan precinct to chat every evening. Groups of domestic maids from nearby countries such as the Philippines also meet in the new town centers, as an alternative to the crowded Orchard Road rendezvous spot. At the town centers, food and shopping are also likely to cost less (Fig. 3.33).

Somewhat like ecological niches, spaces are carved out of the new towns by recurring spatial practices, invoking new spatial practices in new public spaces.

Re-visiting the mapping of daily routines, it is possible to draw out temporal zones of the day when the greatest likelihood of meeting in public places take place, and consider these as "public time". Such a concept helps to explain the rhythmic flux in the liveliness of these spaces, which may be quiet at certain times of the day, but become lively animated spaces in "public time". A glimpse of the sentiments of HDB dwellers about how they use, occupy or feel about new town spaces are presented here in unedited form[82]:

Kumar Sam, male, Indian, Salesperson, age 28, 4-roomer, "I use the public spaces downstairs to meet my friends".

Mdm. Lim, elderly female, Chinese housewife, 4-roomer, "I go down only when there are activities going on, like senior citizen's programs, or when the water fountain is turned on".

Mdm. Chan, middle-aged Chinese housewife, 4-roomer, "I don't use the void deck because it is always occupied by old men".

Anupriya, Indian student, 4-roomer, "I meet my friends at more centralized areas, so that's normally outside of the precinct".

Boon Ping, Chinese Manager, age 29, 4-roomer, "I normally meet my neighbors at the lift lobby or in the lift".

Marion, middle-aged Eurasian teacher, "I prefer to go to the swimming pool".

Thomas Lim, Chinese student, 5-roomer, "I don't use the spaces downstairs, but I think the old folks do".

Winnie Poh, female, Chinese, homemaker, aged 49, 4-roomer, "They use the void decks under blocks 631 and 635 for weddings and funerals. Very noisy, lah! Children cycling all over the place – that's dangerous for us. I also don't like the teenagers using skateboards, too. That is why I rather go somewhere else".

Mohammad Adril, male, Malay, Manager, aged 46, 4-roomer, "I feel that the corridors are too narrow for interaction – maybe they should be wider".

James Yip, male, Chinese, student, aged 14, 4-roomer, "I meet my neighbors at the basketball court".

Yap Lay Hua, female, Chinese, homemaker, aged 28, 4-roomer, "Skating and scooter biking is dangerous for other users of the space downstairs. Besides, it's not so convenient to go downstairs. Corridors are where I meet with most of my neighbors".

Rosnani, female, Malay, homemaker, aged 48, 4-roomer, "" I use the space right outside my flat".

Abu Bakar, male, Indian-Muslim, retiree, aged 58, 4-roomer, "I used the Multi-purpose Hall for my daughter's wedding".

Chong Ah Lan, female, Chinese, homemaker, aged 64, "Older estates like Ang Mo Kio have more facilities. Here, I only take my grandchildren to the playground".

---

[82]These anecdotal comments are collected as part of a research in which I was the principal investigator, *Design, use and social significance of public space in public housing*. The study was focused more on the precinct space than other types of new town spaces.

Within the macro gestures embedded in planned new town structures, the emergent and temporal spatial practices of everyday life enacted by flat dwellers reproduce space in the image of the Singaporean communal. These are the emergent new practices that inform the re-contextualization of new town culture and social systems, situating spatial practice in concrete spaces, bringing to real terms the generalizations of use conceived in the town planner's eye.

3.4.6 *Defiant Spatial Practices*

Just as social spaces and niches are found by New Town dwellers, there are many for whom living in public housing had been an alienating experience. Filmic, poetic and fictional narratives set in public housing tend to offer the following perspective: "(t)he displacement of individuals and their rootedness in Singapore, in personal histories, circumstances and the environment…crops up again and again…its presence seems to be intended to alert readers and audiences to the gritty and sometimes disturbing reality behind the façade of uniform housing" (Yeo 2004, p. 24). These aspects of public housing seldom get an airing in the official discourse of public housing success, or even in local, mainstream television sitcoms such as the happy, multiracial neighbors portrayed in "Under One Roof".

Portrayal of HDB life often takes place in generic HDB spaces such as void decks, corridors, or flat interiors which give a sense of placeless-ness, and also represents the alternative of the global city, as exemplified by places like Orchard Road or the Central Business District. Discomfort and a sense of alienation can be read in another of Sa'at's blog entries. This entry, titled "Railings" (2004), is about something as innocuous as railings in an HDB block of flats:

> As a child, I was always warned by my mother not to touch the railings that zig- zag vertically on the medial side of HDB staircases. Her rationale was that 'people spit from above', and consequently, the metal railings are coated with saliva, contaminated and thus stripped of its supportive function. They become, instead, sites of danger, where initially they would provide some safety, especially if the stairs are slippery.
>
> Another site which represents a potential space of hazard is the threshold at the periphery of the block, where one walks from the void deck into 'open air'. This is a space of vulnerability, where one is exposed to infelicitous weather: a rain of tissue paper, cotton buds, sweet wrappers, cigarette ash, and if of sufficient mass, what has been called 'killer litter'. Thus HDB existence is characterized by aerial threats, where a mouth expels its contents, a hand dispatches its cargo, before withdrawing back into a planar anonymity (which one hides the culprit, one wonders, looking up), where a primary accomplice to misfortune is gravity.

Cases of killer litter, as the potentially deadly aerial projectiles are called, are frequent in the everyday life of HDB living and impact on the perception of public space 'downstairs' as dangerous places. The deviant behavior of the protagonists of killer litter demonstrate the subaltern culture of HDB dwelling in these instances of the deliberate flouting of rules and vandalism of public space. Other instances of what is deemed anti-social behavior by the authorities that are punishable by the law include urinating in lifts and public spaces,[83] scrawling on public surfaces, etc. Consequently, the HDB features as a panoptic space where residents are constantly surveyed, or under the impression of 'being watched'. Bulletin boards are tacked with warning signs, police contact information, and calls to volunteer for the resident neighbourhood watch team. Lifts are fitted with surveillance cameras that are only accessible to the respective Town Councils. This constant surveillance has contributed to a feeling of alienation and emptiness, which is effectively captured in the last part of Saat's poem on "Void Decks":

> Eyes reveal a meeting-point
> For loners and loiterers:
> A sense of things reduced-
> Conversations that trickle through
> Brief noddings at lift landings,
> Teenage rhetoric scrawled, in liquid paper,
> On the stone-table chessboard,
> (Where the king used to sit)
> The grandiose house-selling dreams of residents
> Compacted in anonymous letterboxes;
> As an afterthought, an old man pees
> Under a public phone.
> A place to be avoided, this,
> How in its vastness it devours hours.
> Little wonder then,
> Why residents rush through void decks
> Back to the cramped comforts of home
> As if in fear of what such open space might do
> To cozy minds.

Acts of defiance against the rules of living in public housing are enacted at many levels. Even children who unwittingly play in void decks, ignoring signs prohibiting games such as football and in-line skates, are defying the rules that govern the void deck space. In early 2016, Town Council officials put up posters at the void deck of a block of flats to ban the playing of chess in common areas due to complaints that the elderly men playing draughts "often made noise" and "obstructed the linkway" (Straits Times 2016). The town council later removed these posters, apologized, and put up revised posters with "more measured wording that simply reminded players to be considerate" (ibid.).

Previously, another town council, in an even more zealous bid to stop children from playing ball in the space, actually nailed barbed wire on the walls of void decks, causing the ire of one reporter:

> …in the heartlands, town councils have hammered nails into the walls of void decks in Housing and Development Board estates and attached barbed wires on crossbeams to stop kids from kicking a ball…the nails and the barbed wire portray a people whose behavior

[83]Surveillance cameras have been installed by the HDB where cases of urinating in lifts are recurrent.

> entrenches us in the Third World. Which First World country would use nails and barbed wire to prevent children from kicking a ball?
>
> When the news broke early this month, the Tampines and Pasir Ris-Punggol town councils had already subjected dozens of HDB void decks to nails and barbed wires. More would have suffered the same fate if not for the bad publicity the news generated. Many wrote to newspapers, expressing concern.
>
> Play is part of growing up. It is essential to learning and development — more so, perhaps, than schooling…It is hard to find open space to play in Singapore. About 90 per cent of the population of 3.8 million is crammed into HDB flats, which are built cheek by jowl with generally little room left for physical activities. The open spaces assigned officially for playing are often crowded, far from the flats and blazing hot in the afternoons (Straits Times 2004).

Although such drastic measures are laughable, serious punishment had been considered time and again by the authorities to deter these practices in public housing (Fig. 3.34).

A more subtle defiant practice that only becomes obvious when recurrently practiced over time and space, is the act of walking. Making one's own way through a new town is a kind of tactic, such as Michel de Certeau would describe: "The act of walking is to the urban system what the speech act is to language or to the statements uttered" (de Certeau 1984, p. 97). De Certeau has painted a vivid picture of how a walker in the city is in the act of becoming rather than being, actualizing possibilities, improvising, transforming, and abandoning spatial elements. Through these choices, the walker creates a unique set of spatial experiences through the city.

Indeed, investigating the act of walking through new town spaces already helps us to construct the broad strategies laid out for such an activity and how these have been observed or subverted in the practice of walking. Explained earlier, in new towns, blocks of flats are laid out with void decks at the ground level to create the impression of openness. Open footpaths link these blocks to other spaces in the new towns and are the prescribed pedestrian thoroughfares. Often, the walker takes a serendipitous route through the new town to get to his destination, ignoring footpaths and cutting through grassed areas, moving under void decks, and across open, hard-paved car parks. In other words, he often transcribes a straight line between where he started and where he is going.

As new towns developed, concepts of separating people from traffic became more concrete, and covered footpaths linked blocks of flats and led to common amenities like bus-shelters, markets and other places. However, the traces of the walker through the new town can still be discerned—it is not uncommon to see well-worn paths cut through grassed areas, diagonally transgressing prescribed covered foot paths. While some of these informal routes are eventually paved over to create new paths, the insistent re-turfing of such worn out patches of grass also take place, with new barricades placed alongside footpaths to prevent straying off onto the grass. At the same time, crossing the street at undesignated spots was deemed 'jaywalking' and susceptible to a small fine if the transgressor is caught in the act. Such conflicts between walkers and the managers of new towns occur on a daily basis, and produce creative 'walking rhetorics' encompassing un-prescribed experiences of the new town.

Fig. 3.34 Defiant practices

The most recent phenomenon in new town public spaces is the practice of *le parcour* or parkour (i.e. free-running),[84] in which perpetrators of the activity move quickly and fluidly through space without pauses. Parkour movements including vaulting and jumping over obstacles like railings and walls. The practice caught on in Singapore rapidly, perhaps due to the challenge it gives adolescents over the banal new town landscape.

Personal rhetorics are also inscribed on the facades of HDB flats through customization of the corridor in front of one's flat to distinguish it from one's neighbors' and also at the back façade of flats, where the 'laundry parade' breaks the monotony of the uniform HDB façade. On the part of the HDB, 'improved' flat designs are constantly being developed, particularly to remove the 'unsightly' sight of laundry by hiding the drying spaces in recessed alcoves, and also by removing as far as possible the long corridor such that segmented corridors with very little frontages are provided (Fig. 3.35).

[84]Le parkour is adapted from the French word *parcours* which means 'by path'. The practice caught on in Singapore after French director, Luc Besson's film, Yamakasi—Les samouraï des temps modernes (2001) became a cult film for urban youth. The film depicted a group of disenfranchised youths from housing estates in France, perhaps stirring a chord of empathy from the highly confined lives of Singapore adolescents.

Fig. 3.35 Parkour in action at HDB estates. *Source* Mark Teo, courtesy of Superfly Monkey Dragons

3.4.7 Spatial Practices in Transition

Such an overlaying of practices in space, accompanied by spatial narratives and anecdotes, helps us to conceive of the dense texture within which public space is used and experienced. The layering of routines and commentaries, capturing the points of views of those who inhabit (or choose not to inhabit) public space, suggests various meanings attached to public space. The spaces of the new town, although prescribed as the public space of the community, often tend to merely accommodate isolated groups who appear within the space but have very little interaction with others. Such spaces seem to suggest the appearance of a community, but this remains on a symbolic level, as individual bodies that inhabit this space mainly perceive others and do not come into social contact or 'friction'.

The public spaces of the precincts represent the community, and their value lie perhaps in that they concretize the territorial space of those who live within its confines. They make visible their members and thus impart knowledge and information about others. The playing out of cultural differences within such a space is not always welcome, such as instances of the Chinese sometimes burning paper offerings in the space, or the Malay tradition of cooking in public, but these practices are nevertheless tolerated. However, the layering of individual narratives over space and time creates texture invisible in the space. Many of the everyday spatial practices of residents leave no trace, but add to the meaning and constructions of public space.

Within the larger context of the new town, a system of channels of nodes constitutes its spatial movement system. These public spaces create ample opportunities for the invention of new spatial practices as well as room for the adaptation of persistent spatial practices within its framework. Although the hierarchical system of the new town structure tends to be rigid and it disperses rather than assemble communities, the increasingly coherent movement system also tends to bring people into more concentrated channels and nodes, allowing an overlapping of routines and space in "public time".

The display of defiance in public space suggests that a narrative of loss exists, drawing from a time and place when and where people were better grounded in society and in space, instead of the rootlessness of moving from one new town to another whenever there is a change of circumstances: i.e. a young couple marry and move to a new flat in a new estate, or in the effort to 'upgrade', a family moves out from their flat into a bigger, newer flat in a new estate. The HDB's upgrading scheme is known as the Selective En-bloc Redevelopment Scheme (SERS) and it allows the HDB to earmark certain estates for redevelopment and to proceed with its plans if the majority of homeowners acquiesce to the sale. In turn, residents receive compensation and right of first refusal in selecting newly built apartments in the vicinity. This is the HDB's way of incentivizing renewal and optimizing land use; however, one of the unintended consequences of this scheme is that it disallows a building and its residents to age in place. The lack of memories about place and the lack of distinctive difference in many of new town public spaces contribute to the abstract nature of their perception and representation. Without psychoanalyzing this phenomenon, what is certain is the divide between representations of public spaces in the city and public spaces in public housing. There is little 'middle ground' in the transition of these spaces, but rather an abrupt and mutually exclusive division. HDB spaces are indeed 'exclusionary', where a belief that a truly Singaporean vernacular exist, and where national identity is cultivated. It is no wonder that the HDB landscape is also often used in justifying the political legitimacy of the ruling party (Chua 1997).

As new town models continue to develop, such as the more grid-like layout of Punggol 21 New Town, the structure of new towns, as well as the forms and types of public spaces they engender, continue to change. The new generation of flat dwellers, who were born into the HDB 'vernacular', knows of no other spatial practice than that within the HDB context, as opposed to the earlier generation of pioneer flat dwellers who brought with them an earlier tradition of spatial practice. This phenomenon is bound to have an impact on the public space of younger new towns, which have constitutionally younger demographics. While some HDB estates are better planned than others, there remain under-utilized amphitheaters, out-of place playgrounds and confusing stairwells in older HDB estates. As Singapore continues its urban tinkering and re-provisioning of its public space, the government should prioritize public participation in the process, in order to ensure new developments coincide with natural linkages and social uses of spaces. The forms of public space and spatial practices in new towns remain in flux and are constantly in a state of transition. Perhaps these too are distinguishing characteristics of new towns that have to be taken into account in understanding their public spaces.

Chapter 4
Themes and Modes of Production of Singapore Public Space

4.1 Constructing Public Space: Themes

Building from historical context and the examples brought up in the three case studies, I have identified themes that have permeated through the construction of Singapore public space. These include themes of identity, place, congested space and compressed time, programs, continuity and change, control, and public space in flux. From these themes, we will construct a framework to help describe the modes, methods, and production of Singapore public space.

4.1.1 Identity: Ethnic Space, Ethnoscape, and (Spaces of) Exchange

It is evident that even as associations of public spaces in ethnic quarters in the city are retained, new identities are being constructed of place and its inhabitants. The histories of public spaces in the case studies, such as Little India and Orchard Road, are being selectively re-imagined and constructed by the state, visitors and inhabitants. Identities in these spaces have not remained static, but have been infused with newly injected bodies, bringing with them new perceptions, practices and meanings to the physical spaces.

4.1.1.1 Ethnic Space

As we can see from the case of Little India, ethnic space functions as the physical repository of the history and culture of an ethnic group but it is also expected to be relevant to contemporary urban life. While such spaces promise the visitor unique experiences of ethnic culture, they also link inhabitants to the material culture of

L. Hee, *Constructing Singapore Public Space*, Advances in 21st Century Human Settlements, DOI 10.1007/978-981-10-2387-3_4

everyday life as the stage on which their everyday ethnic life practices are constantly re-enacted. The notion of the ethnic community is constantly changing. The definition of community has historically constituted the people who live within an area and share common ethnic roots as well as resources. But the communities today who share the ethnic public space may come from geographically dispersed networks. These communities now find a common identity only within the symbolic dimension of ethnic public space.

Cultural practices, decorum and rituals may be enacted by members of communities within public space to re-create their common identity. With their local knowledge of the ethnic space, they produce the notions of "insiders" and "outsiders". These definitions of identity are fluid, as groups or individuals may tend towards adopting multiple subject positions and can be either insiders or outsiders at different times, depending on the situation. I have defined the field of spaces within ethnic public space as "front" and "back" spaces: the front spaces being where the definitive cultural and spatial practices of the ethnic group are enacted, while the back spaces are where practices that are against cultural and behavioral norms, or are on the fringe of public acceptability, are enacted. Here, ethnic public space not only serves as the physical repository of spatial practices and culture, it is also a contested place where different groups with varied (and possibly conflicting) interests are brought into proximity. Thus, ethnic space may not be a homogenous space—it embodies contested territories and zones of particularity.

The extension of cultural practices into public space is not only confined to outwardly ethnic public spaces, but also occurs in the HDB heartlands of Singapore. Private or familial events such as weddings, funerals and birthdays, as well as modes of worship including the burning of 'paper money' by the Chinese, or the Indian practice of burning incense, are exteriorized through their enactments in public space. While these practices allow Singaporeans to gain a better understanding of the multiracial composition of the larger HDB community, they also highlight cultural differences and at times, cases of intolerance and conflict. However, with the replacement of the older generations of flat dwellers with younger generations of New Towners, these practices are in flux.

4.1.1.2 Ethnoscape

Adding yet another dimension to public space is the globalized "ethnoscape". Singapore, with its increasing reliance on global labor, has continued to attract large numbers of 'expatriate' professionals and lowly-paid 'foreign workers.' The emerging and highly visible "ethnoscapes" of these international subjects attest to the strong impact of globalizing economic and cultural processes within the cityscape, as well as the resulting social and spatial antagonisms that arise. The definitions of the 'global' as universalizing forces, and the 'local' as endemic forces resisting them do not apply in both the cases of Little India and Orchard Road. On the contrary, some of the global actors within these public spaces (i.e. the foreign workers) introduce spatial practices that could be described as traditional regionally derived cultural practices that the local actors may have moved on from. These foreign workers regard public space as a social resource to meet the needs of companionship, the sharing of information, entertainment and identity building—in other words, they see public space as spaces of exchange. The visibility of this 'foreign worker ethnoscape' within prominent Singaporean public space heightens the locals' awareness of cultures that are different from their own, as well as the plight of these workers within the context of their working and living conditions in Singapore.

The visibility of other cultures in public space initiates the redefinition of Singaporean public space and spatial practices. The cases of Little India and Orchard Road reflect spaces of exchange functioning like portals through which ideas, people and notions of public space flow and overlap. As a space of exchange, Orchard Road brings Singaporeans of different backgrounds to meet, although not every strata of the Singaporean population is represented here. The ethnoscape in Singapore public space is less analogous to a 'tossed salad' than to a 'clumpy soup,' constituted by different groups aggregated within different zones, as illustrated both in the examples of Little India and Orchard Road. The phenomenon of the Sunday enclaves is bounded by space, in time and overtly, by social distance. Discussions in the local media regarding this phenomenon acknowledge the existence of the foreign worker groups and foreground their problems, giving a public 'voice' to these otherwise 'silent' groups who are marginalized in social and urban life.

4.1.1.3 (Spaces of) Exchange

The nature of public space at Orchard Road is that which different social spheres simultaneously exist in proximity to each other, instead of them sharing and using the same space. The value of Orchard Road as the public space of the nation lies not so much in it being a shared space, let alone a true 'meeting' of different spheres, but in its provision of opportunities arising from urban proximity. More than any other space in Singapore, Orchard Road is a space of friction. Here, ideas of the self and of others are constantly evolving through dynamic negotiations of space between different groups. Orchard Road allows one to experience otherness through an exposure to spatial practices and views other than one's own. It enables possible 'shifts of perspectives' that can bring about pleasant or antagonistic (even offensive) exchanges.

Little India, as the site of overlapping interaction between the spheres of ethnicized and globalized public space, provides a common ground for the discovery of similarities and differences between groups and communities. These similarities and differences tend to become more apparent in particular "front" and "back" public spaces, if one can read the indicative signs and signifiers legibly. These environments are not always defined as demarcated urban configurations; instead, they are often regarded as overlapping spatial fields. While front spaces tend to be well defined as contained or prescribed spatial vignettes, back spaces are often embedded in dense networks of connected spaces. The overlapping networks of public spaces in the locality allow for routes to cross. Meetings, whether routine or serendipitous, do occur in public space, and users and inhabitants through their everyday lives adopt these encounters as the social and imaginative space of the community.

'Front' public spaces in Little India tend to coincide with public spaces that women inhabit. But women remain the passive consumers of public space here—their presence and identity become constructed and defined by the spatial practices they engage in their everyday life. To some extent, the notion of "agoraphobia" might apply to them (Deutsche 1996, p. 325). Agoraphobia is a kind of panicked reaction to the openness of public space, partly due to the impact of social restrictions to which these women are usually subjected: "In city streets and squares, where men have greater rights, women devise strategies to avoid the threats that present themselves in public spaces. The phobic woman may try to define, and stay within, what she considers a zone of safety" (ibid.). The spatial practices of women in these ethnic public spaces play out through deep, embedded social structures that are made visible in these gendered public space.

Not all public spaces serving a 'community' are necessarily spaces of exchange. Public spaces in the HDB new town tend to hold segregated groups who inhabit the space. In the context of these spaces, the notion of community is mostly symbolic. Individual residents merely perceive others, and may not come into social contact with each other, or experience 'friction.' The value of these public spaces lies in the fact that they represent the community, and make its members visible to each other within their spatial boundaries. These spaces allow their inhabitants to know more about others living within the space. However, these inhabitants might not appreciate the actual cultural differences played out in these public spaces.

4.1.2 Place: Non-place, "-Scape," and Images of Elsewhere

4.1.2.1 Non-place

Spaces which are considered Augéan non-places—places to pass through and which have no associative meanings—when considered through the perspectives of those who actually use them, become places with the potential for unique experiences. These places are often 'bottleneck' spaces such as subway stations, underground crossings and bus stations, in which people with different spatial itineraries are brought momentarily into close proximity. These are also places of possible encounters as they are well connected with networks of public places, where one may meet and re-define Others. As places, these are not always alienating environments. Often, movable 'fixtures' such as buskers, vendors, or even people who use these spaces at regular times, such as those commuting on regular workdays, form familiar entities within these spaces.

Although void decks and public corridors in public housing new towns seem to function as non-places within such environments due to their nondescript aesthetics and placeless-ness, they are often the site of encounters and where otherwise private events such as weddings and funerals take place. As these spaces are inadvertently part of the system of spaces through which one transits from home to the outside world, they are similar to the earlier 'non-places' of transit spaces in the city. But these generic spaces favored in portrayals of the visible everyday life of HDB dwellers in the media are weaved into the everyday life of new town dwellers, such that they form an integral part of daily routines within estates. Though placeless in character, the void decks and corridors have become spaces of public significance, occasionally becoming the extended space of significant private events.

4.1.2.2 "-Scapes"

"-Scapes," or nodes of public space dominated by particular groups, feature characteristic forms of spatial practices that feed off the social structures and milieu of the times and places in which they exist. "-Scapes" are temporal in nature, and occur where transient conditions are favorable—the youth-scape of Orchard Road that emerges on Friday nights is one such example. The behavior of "-scapes" is sometimes dependent on the technology that supports the spatial practices and habits of groups, such as the mobile phone, Internet, and messaging applications. The level of visibility offered by the public space of Orchard Road, the perception of what is 'hip' and 'happening,' the mutability of social networks, and other subjective conditions make these "-scapes" difficult to anticipate.

"-Scapes" hold social and, sometimes, symbolic significance for their inhabitants, and their spatial field is constructed to define insiders and outsiders by discernable codes of behavior, age-group, gender, nationality, or local knowledge of the district. "Localities," such as those discussed in the Little India case study, are defined within public space and are embedded within the imagination of their inhabitants, setting the inhabitants apart from the outsiders. The locality, however, is not a homogenous entity. Within the field of localities, an even more deeply embedded field of 'public' spaces exists, situated in the ambiguous space between public and private, legitimate and illegitimate. In the case of Little India, the grey-scape is the Red-Light district harboring several types of fringe activities. On Orchard Road and in Little India, the Red-light districts are obvious only to those who have local knowledge of the area, or can decipher the signs and signifiers leading to the subverted spaces. Accessibility to these grey-scapes are determined by the physical enclosure of these spaces, the degree of visual control one has over the space, the thresholds of interiority which one has to penetrate to access these private realms, the gender-based rules of operation, and the selectivity of participants who have access to these spaces.

Women are ubiquitously absent or not overtly present in the subverted "-scapes" or 'back' spaces. Instead, women in public space often seek legitimate reasons to be in public, and the practice of shopping, such as on Orchard Road and in Little India, provides the ideal alibi to be in public space. The infrastructure of shopping is especially well-provided on Orchard Road, where a whole plethora of activities, services and spaces supporting shopping, such as eating places, public exhibitions, personal services, rest stops, public restrooms, and various forms of transport and transit spaces, can be found.

4.1.2.3 Images of Elsewhere

An extreme form of "-scapes" is the import of images from elsewhere into Singaporean public space, where these images subsequently take on a life of their own. The transposed images of Christmas in distant lands, with the accompanying visual effects of snow, the Northern Lights, and traditional symbols of winter holidays, like fir trees, angels, bells and snowmen, transform the public space of Orchard Road to create new experiences and new nodes of activities for its visitors every end of the year. As a sustained event for the national collective, the dis-embedding of foreign imagery and re-embedding of it into Singapore public culture has transformed the hyper-real into real notions of 'event' and 'festivity' in public space.

Although ethnic public spaces in Little India have developed historically in their own right, they are superimposed with images of places from elsewhere such as Madras—sans noise, dirt and smell. The images are packaged in tourist promotion brochures to be consumed as spatial vignettes or tourist-scapes: they range from the covered shopping streets of Little India Arcade to festive glitter of the Deepavali light-up for a month every year. The inhabitants and culture of the quarter are packaged as a homogenous entity (i.e. the Indian ethnic quarter of Singapore) and its public space is set as a stage for the re-enactment of cultural practices and customs. These tourist-scapes are often promoted as the definitive "front" public spaces of the district.

There have been attempts by the state to generate festivity with mass events held on Orchard Road. These mass events lack the spontaneity associated with urban street events—they usually feature coordinated movements and rehearsed steps with participants 'following the leader' with mimicked dressing and dance steps. There is a strict division between performers and audience, with performers enacting staged actions in front of a sanctioned crowd. While the importing of Christmas imagery into Singapore public space has become an Orchard Road tradition, other activities such as street dancing died down after a while due to a lack of public interest. The Singaporean public is not spontaneous and spirited enough to sustain such practices.

4.1.3 Congested Space and Compressed Time

Due to inadvertent congestion in the limited space of the city, public space tends to be contested in both the dimensions of space and time. The phenomenon of congestion creates both liveliness and antagonism in public space. Tensions of physical space are engendered in what I referred to as "threshold spaces," in which public space is extended physically through its claim on private spaces or vice versa, through the use of various devices including movable objects or other signifiers of claimed space, or through spatial practices. These tensions created in threshold spaces include conflicts between ownership, modes of traffic, and angst due to the congestion. However, these contested spaces also witness lively interactions and conviviality, as people are brought into congested zones or into close proximities, increasing chances of social contact or 'friction.'

The claim and appropriation of spaces within temporally defined zones are exemplified by the public phenomenon of the Sunday enclaves, which can be found in several spots in Singapore and illustrated within the case studies of Little India and Orchard Road. These practices lend alternative interpretations of how public space can be used, and they re-define what constitutes public space. The temporary appropriation of all kinds of open spaces imposes an entire system of spatial practice over the existing spaces, engendering a whole new topography of public spaces. Public spaces of congestion not only present themselves as sites of spatial improvisation, but also as sites where all the groups using the space become visible. New meanings and spatial practices are borne out of the intense appropriation of space, which sometimes yield lingering influences and traces of the appropriating groups within the public space outside of its temporal boundaries.

Other forms of public spaces occurring within overlapping concentrations of space and time include void decks, paths, neighborhood coffeeshops, and the vicinities of markets, food centers, and transit stations. These form channels of activities that are interspersed with nodes, constituting a spatial system in which recurring spatial practices that are enacted on an everyday basis become embedded. The configuration of new town centers, with converging routes along the shop fronts, create intensely used paths. These paths are where contestations of space between retail and public thoroughfare are enacted, reminiscent of the contested five-footways of shophouses in the city. The channels formed through the orchestrated routes in the town center increase the chance encounters of neighbors and acquaintances, thus becoming social spaces. While the rigid structure of the new town usually disperses rather than assembles communities, these concentrated channels of activities and nodes allow for a system of movement of "public time," with peoples' routines overlapping within peak temporal zones.

The framing of public spaces within temporal zones is exemplified by spatial practices along Orchard Road, in which 'public events' are staged on cyclical, regular or ad hoc bases. The public spaces are often transformed to some degree during these events, depending on the scale of the event, and these transformations are bracketed in time. The high visibility of a main street like Orchard Road makes it the ideal public event space. These events are not limited by the timescale of their staging, as they are broadcasted and transmitted through mass media, extending the spatial and temporal boundaries of their existence.

4.1.4 Programs: Programmed Space, Scale of Engagement, and Choice

One feature of the lively public spaces in the case studies is that these spaces often have flexible or overlapping programs, offering a variety of choices on how they can be used and appreciated at different scales of engagement—from the personal to the urban. The potential reconciliation of different programs and scales enriches the fabric of public space and enables differing modes of engagement with public space. The perception of public space from different scales and from multiple subject perspectives is possible only when physical space embodies such a capacity, and is enabled by programming.

For example, Orchard Road presents a myriad of sectional profiles along its length, allowing for different types of relationships between people, buildings and the street, as well as a whole range of interiors, in-between and exterior environments that one can inhabit. The lack of uniformity on the Orchard Road frontage gives the street a different character from many European shopping streets. There is a secondary economy of street-side kiosks and an informal economy of vendors selling items, ranging from ice-cream to drawn caricatures. Although the urban fabric of Little India seems to conform to the shophouse block typology, the array of social spaces found and improvised from the available spatial template is surprising. Within public housing, it is often not the highly programmed spaces that are lively and convivial, but those that allow for different programs and flexibilities of use.

An entirely different view of public space may be perceived and understood at the personal level, as opposed to the level of urban phenomena, or from the official rhetoric of public space. While Orchard Road is commonly understood as a place of leisure, people who work within Orchard Road also have an impact on these spaces. Vendors and street buskers may serve as semi-permanent fixtures in a space. They define the spaces around them and function as reference points or time-markers within public space. Street cleaners after a big mass event restore a sense of the everyday back to the space.

Spaces vary in the degree of programming and some degree of improvisation occurs in their use—spaces range from highly programmed departmental store atrium spaces to incidental spaces that act as a buffer zone between the public and the private. In Orchard Road, arcaded spaces lining the buildings on the street create zones that provide public relief from the tropical heat. Likewise, the five-foot ways of Little India act as in-between spaces, for resting, waiting and meeting friends. These 'big-city rooms' or intermediate zones form part of the network of 'plug-in public spaces' and have a life of their own.

Due to the wide range of spatial configurations offered, a public space like Orchard Road enables one to assemble one's own itinerary of public spaces. Little India works in a similar fashion, but with more densely overlapping networks of social spaces, depending on one's position as an "insider" or "outsider". The public housing spaces are connected through a system of transitory spaces, which present opportunities for unexpected social encounters. While these spaces differ in the degree of social mixture, from ethnoscapes to ethnic spaces to community spaces of the "heartland," they allow a constantly evolving identity of self and non-selves to be formed.

4.1.5 Continuity and Change: Everyday Life Space

The spaces of everyday life in the HDB new towns exhibit a wide range of re-contextualized spatial practices. The spatial practice of extending private property into public space had persisted since the times of the communal *kampongs* of the Malays and Chinese, as well as the five-foot ways of shophouse blocks, and the claiming of the streets for various functions of living. The celebration of significant private events in public common spaces has become part of a HDB 'void deck culture.' These spatial practices that extend the private into the public also apply to commercial activities in the new towns. Configurations of physical spaces and their relationship to the home in Singapore may have changed, but the Singaporean notion of claiming public space for private use continues.

New spatial configurations of new towns have also encouraged changes or transformations of spatial practices into new forms that nonetheless retain their former social meaning. New networks and shapes of physical spaces present new opportunities for people to inhabit them or for social encounters to occur. The convergent paths and routes in the new towns leading to common centers of commerce and recreation help to increase chances of neighbors meeting. New meeting spaces have also emerged, such as in the vicinities of the new town shopping malls, coffeeshops and transit centers. Different groups have also discovered their own niches. These niches seem to operate on an ecological logic of their own, often constituted of particular demographic groups who meet at regular times of the day or week. Within new towns, the meeting of paths and daily routines create concentrated zones within "public time," in which one is most likely to encounter neighbors within the concentrated nodes and paths of the new towns, or in favorite spots such as coffee shops or at the market.

Although the movement system leading from the apartment blocks to the neighborhood center and town center changes from linear, hierarchically defined spaces to a more distributed system of spaces, the punctuation of nodes within these routes enable a concentration of activities instead of these nodes being just transitory places. These emerging spatial practices of everyday life, seen in the context of the new towns as a whole, shape a network of social public spaces that develops over time and becomes entrenched within new town social culture.

The undifferentiated and regulated nature of public spaces of HDB new towns can elicit feelings of discomfort and alienation in their residents. Spatial practices that defy rules of public housing spaces, or transcribe trajectories within space not anticipated or sanctioned for these contexts are sometimes attempts to leave a personal mark within space regarded as anonymous or alienating. These may include acts of claiming space that include residents leaving personal items or furniture in public space—such as groups of elderly men who assemble their own 'living room' within void decks. Other defiant or subversive acts include leaving traces in public space—such as graffiti, vandalism, littering, and the throwing "killer litter"—as well as acts of crime. Another recurring, subtle form of defiant practice in public space is the act of walking—not following the neatly laid out system of paths in the new towns, but making one's way serendipitously through all kinds of ground surfaces. One inscribes a personal rhetoric within spaces of the new town by creating new paths through turf areas and disregarding existing footpaths.

The range of defiant spatial practices exhibits creativity, a need for identity, inconsiderate behavior, the venting of frustrations, or reflect criminal behavior. Another phenomenon in new town public spaces is the practice of *le parcour* or parkour (free-running) in which perpetrators of the activity move quickly and fluidly through space without pauses by vaulting and jumping over obstacles like railings and walls. The practice of parkour is a way for adolescents to transform their mundane experiences of new town public space through acts of improvisation.

4.1.6 Control: Negotiated Space

As discussed in the earlier chapters, the state is determined to control the behavior of its citizens in public space, as any form of organized protest or chaos in public may signal impending dissent. Assemblies of over five persons in public require a permit, and rules govern the practice of 'free speech' at the Speakers' Corner. Control can be passively enacted in many ways, such as through the design of spatial form, and in the omission of large spaces of assembly. It can also be overtly enacted through forms of spatial enclosure that limit or increase degrees of visibility and accessibility for the user; and also through gates and barricades, rules and regulations, and various forms of surveillance. Active forms of control mainly involve police action.

The case of Little India illustrates many examples of improvised spatial practice by the foreign workers in which rules of 'correct' behavior in Singaporean public space are broken. Foreign workers engage in practices such as sleeping in public space, flouting traffic rules for pedestrians, occasional brawls, obstructing paths and loitering within the void decks of flats in the area. In response to these practices, passive forms of control such as barricades have been erected along paths and around void decks to prevent undesired loitering, and surveillance cameras have been installed in the area. Police gantries are set up on Sunday evenings for crowd control along the main street. Although the police strategy might at first seem strong-armed, a closer look reveals that the control exerted by the police here pertains more to the apportioning of rights and the managing of conflicts over the use of space. The stance of the police has been one of pragmatism and improvisation in managing not just the bodies of the workers in space, but also the occupation of the workers' time with activities such as ball games, public movie screenings and a flea market. The spatial fields occupied by the workers have also been shifted out of the congested streets of Little India into the open spaces across the main street.

Space has to be constantly contested and spatial practices negotiated as spatial practices are inexorably bounded with restrictions, especially for marginalized groups of foreign workers. In the case of the Indian migrant workers, they had managed to find open spaces to lounge or to conduct cricket matches in their free time. As for the Filipino Sunday enclaves on Orchard Road, the limits of private building regulations are tested as these foreign workers occupy both public and private spaces. The Filipino workers' success in appropriating the space have somewhat emboldened them to stake their claims elsewhere in the vicinity and also in the new towns. The phenomenon of the Sunday enclaves in public space has pushed the envelope of tolerance of Singaporeans and the state into a form of "pragmatic benevolence".

The persistence of the Sunday enclaves and the need to deal with their presence has transformed public space and spatial practices. A new spatial practice of dense congregation enacted by the foreign workers which embodies the logic of a seemingly chaotic crowd, has engendered new forms of negotiation between the enclaves and those controlling the space. New devices for containing the crowd in public space have emerged. These involve the dispersal, negotiation and minimizing of conflict, as well as the mediation of differences. Negotiated thresholds of publicity often transform these spaces into sites of tension, or intimate friction.

The visibility of the foreign workers in public space has played an important role in heightening Singaporean awareness of social problems associated with imported labor. The cases of Little India and Orchard Road have also demonstrated a coming-of-age of a dormant civil society, in the form of Singaporeans who have stood up and spoken for marginalized groups like the foreign workers, and non-governmental organizations who have provided some form of support and relief for these groups, such as free meals, the provision of forms of recreation and entertainment, and making available affordable goods. The Singaporean social landscape is dynamic and changing, reflecting the practice of negotiation through practices within public space.

Levels of electronic surveillance in Singapore public spaces have increased—as is the trend in many other cities in the world, such as in London and Rome. Cameras have been installed at various public places in the city, as well as in new town public spaces, as devices to prevent crime and vandalism. Contrary to the reactions of protests in cities in the West, Singaporeans in general regard surveillance as a necessary measure and have welcomed the additional layer of security. The right to individual privacy in Singapore is often subsumed under the establishment of collective security for the benefit of the greater good. In its most extreme case, the form of electronic surveillance for an individual may begin the moment he steps out of his flat into the common corridor, through the multi-storey carpark of his neighborhood, through the electronically monitored road gantries, and into the lobby of his office, the bank or departmental store.

The form of new towns, with its increasingly contained clusters or precincts, tends to compartmentalize space beyond the HDB's original efforts to disperse close-knit *kampongs*. As precincts are visually closed in and blocks focus into the common open space, the activities within the precinct space are well surveyed by eyes from above. The degree of privacy of the precinct spaces has increased dramatically since the early days of public housing, due to adjustments in building configuration. The highly prescribed nature of the usage of these spaces does not make them the liveliest spaces in the new towns.

The monitoring of the precincts is not limited to residents who live within them, but is also undertaken by the town councils, as the cleanliness of public space in the precincts becomes a criterion to determine if the precinct 'deserves' to be a candidate for improvement works. The voting patterns of a precinct during the general elections are also determining factors in its chances for upgrading. Changes to the design of HDB flats also make them less likely places to enact practices of 'personalization.' New design features include the provision of alcoves to hide laundry, the use of tiled surfaces for common areas to prevent graffiti, and the elimination of common corridors in front of units.

4.1.7 Public Space in Flux

The range, types, and occurrences in the landscape of Singapore public space are very much related to the state and pace of development of the nation, which are determined by factors such as economic and commercial developments, impacts of globalization, the availability of land resources, work and leisure culture, public housing development and ideology, everyday life habits, the state of emerging civil society and the degree of control exerted by the authorities over the form and culture of public space.

The fast pace of material and societal change seems to promote changing societal trends and the emergence of new spatial practices. But at the same time, there is a throwback to persistent cultural practices and a sense of nostalgia regarding lost places and practices. While the city-state aspires towards becoming a "Global City of the 21st Century," ethnic public spaces are being preserved or re-invented in the city. Just as new types of simulated public space emerge in the city, such as the annual Christmas landscape, so do choreographed public events involving mass participation in public space. The speed of urban life and the means of transit in the city have seen a proliferation of spaces of transit such that these become places of opportunistic encounters and experience. The speed of transit also means that new modes of connectivity now bring together different networks of public spaces in the city seamlessly.

With the rapidly changing nature of urban space and the range of typologies of public spaces engendered, there is a parallel increase of various types and scales of engagement of the different types of publics with public space. Emerging new sociabilities in the city create a mosaic of "-scapes"—e.g. "youthscapes," "ethnoscape," "kaffee klatsch," "perches," "maidscapes," etc. Terms of engagement with public space involve a constant negotiation of boundaries that shift between public and private spaces, and range from personal experience to mass urban phenomena.

The presence of global entities, including high-end global workers, tourists, low-wage foreign workers, and the local, middle-class mainstream lead to new and observable phenomena in public space. These phenomena include the temporally bounded enclaves such as the Sunday enclaves of Filipinas on Orchard Road and south Indian workers in Little India, which demonstrate how the visibility of different groups in public space offer opportunities for 'friction' to occur. The presence of the state in public spaces is exemplified in different ways, through both active and passive means. Public spaces in the city also have physical boundaries, or they may be intangible or temporally defined. Public spaces within a locale often also have self-defining notions of "front" and "back," "insiders" and "outsiders," and varying degrees of acceptance of the practices of others.

Public spaces in the new towns, being the invented public space of the nation, present an interesting variety of spatial practices. These range from practices that have retained much of their form from precedent spaces, persistent practices that fit into the new spaces as well as emerging new practices derived seemingly from everyday life in the new town landscape. Yet, defiant spatial practices, which may be creative or destructive, are also forms of reacting to the often alienating and indifferent landscape.

With the increasing complexities of urban life and congestion of space, the concept of "public time" becomes increasingly significant over "public space". As less time is spent in public space, recurrent and repetitive practices that take place in public for quotidian purposes, become more likely the basis of change than singularly transformative events. The body acting in public space becomes the transformative subject—not so much in the form of protest than in its persistence of practices.

4.2 Modes, Methods of Production and Representation of Public Space

The current form of public space is characterized by emerging diversity and allows for varying degrees of freedom in terms of spatial practices. Spatial practices reproduce or transform public space over time, embodying historical and everyday processes of growth and urban change. The mode of public space is not limited to its physical form. It also includes spaces that are defined by intangible boundaries or are temporally defined, and spaces that are separated by distance but are similar in nature, as well as spaces in states of flux.

The production of space referred to here does not particularly pertain to processes of capitalism and social relations as raised by Henri Lefebvre, but refers to the sense of "practiced space," i.e. space embodying certain forms of spatial practice.[1] The "method of production" referred to here describes how such a "space," whether physical or practice-based, is defined and continues to be defined on a recurring basis—physical space being a pre-existing requisite for spatial practice to occur. As such, physical space is re-produced as a recognizable "practice-space".

The notion of representation in public space is taken in the liberal Arendtian sense that alludes to political space—'being in public' means being in the 'space of appearances.' Groups represented in public can use public space to address their needs or state a point of view, i.e. be seen and heard and thus have rights. Conversely the absence of individuals or groups in public space would mean that the lack of representation in the public realm negates their ability to have rights.

I will now provide a framework under which the dynamics of these spatial practices (i.e. their mode and methods of production), how they define contemporary public space, and how the public is represented in space will be discussed. Change is often achieved by resisting or transcending the forms of control the state enacts in the production of public space in Singapore. Situating the themes from the case studies with respect to the notions of change and control will help provide us with a useful overview of the state of Singapore public space.

The diagram or map below sets the various themes along the axes of "extent of change" and "degree of control" (Fig. 4.1) and shows four quadrants of spatial modal frames. The axis of "control" moves towards "free-will", while the axis of "change" has at its least extent "preservation". The axes intersect in the middle of the diagram, so that the extremities of the four quadrants represent the definitive modes of "traditional" space, "transferred" space, "transgressed" space and "transformed" space.[2] Each of these modal frames will be discussed in turn, and within each frame, I will focus on the methods of production of public space through spatial practice, and also the notion of representation within these spaces. The elements of phenomena in space and the spatial containers themselves are conflated in the graphical representation and discussed on the same plane of reference. These elements have different degrees of impact on its milieu—for example, the existence of the Sunday enclaves are more impactful on Singapore public space than the phenomena of flash mobs, which exist for no particular reason other than as synchronized oddities in public space. As such, the graphical representations of significant practices or spaces are clouded in white, while those that are less impactful are only represented in text.

---

[1]This is by itself a move for which many observers of space are inspired by Lefebvre's *The Production of Space* (1991), in which space is not to be viewed only as a container of activities, but as a product of spatial practice. The discussions here conflate space and spatial practice so that phenomenon in space and the container are discussed in the same plane of reference.

[2]It may be of interest to note that the way of using terms starting with the same letters, here "tra-" is a particularly Singaporean habit! Other examples include the 5Cs, the 3Rs, etc.

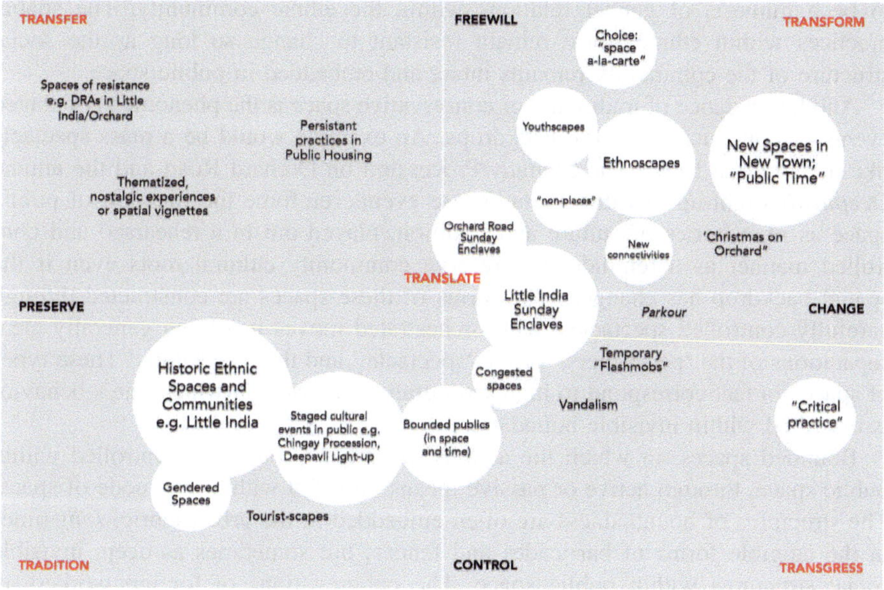

Fig. 4.1 Graphical representation of spatial transitions in public space

4.2.1 Traditional Space

These are spaces in which the form of spatial practice remains largely unchanged over time, with some form of control usually exerted within these public spaces to maintain the status quo. These spaces include what I have referred to as 'ethnic space,' which act as the physical repository of cultural spatial practice and is the space of cultural representation. Although many forces of change may impinge on such spaces, the social structures and dense networks of social relations within these spaces produce and maintain the form of the ethnic space. Physical ethnic space may pre-exist as the space of ethnic spatial practice, but these ethnic practices are overlaid onto space and have their own fields of meaning, which may not be anticipated through the form of physical space.

The case of Little India demonstrates such an example best, as cultural practice is concentrated within space. Although Little India as a district is conserved and efforts had been made to revitalize it with traditional bazaar environments like the Little India Arcade, the best instances of ethnic public spaces do not occur in these gentrified developments, but in 'naturalistic' settings like the five-footway and backlanes, which are by no means spatial forms particular to the Indians. The Indian community, although regarded as a homogenous entity by 'outsiders,' presents itself as composed of heterogeneous entities within ethnic space. The male gender is noticeably dominant, while women and children are absent except in the most 'public' of these spaces. Representations of gender in public space are highly likely to be a mimesis of gender relations within the ethnic community. The spatial practices within ethnic space remain resistant to change so long as the social structure of the community remains intact and embedded in public space.

Another instance of traditional or conservative space is the phenomena of staged events or constructed cultural backdrops. An example would be a mass spectacle like the Chinese New Year *Chingay* Procession on Orchard Road and the annual *Deepavali* Light-up in Little India. These events reinforce the function of public space as repositories of culture and tradition, played out in a rehearsed and controlled manner as a reminder to preserve community cultural roots even if the spatial backdrop has changed. The forms of these spaces are constructed through carefully controlled spectacles, with orchestrated moves involving generally clear separations of the 'performers,' or the 'spectacle,' and the 'spectators.' These types of spaces in fact correspond to highly programmed space, in which one's behavior is inscribed within invisible boundaries.

Bounded spaces, in which the actions of certain groups are controlled within public space, through active or passive means, also fall within this mode of space. The structures of boundedness are often embedded in the urban fabric, sometimes in the tangible forms of barricades and fences, but sometimes as deep, invisible social structures within public space. The congregations of foreign workers in Singapore public space are inscribed within invisible boundaries. Their presence in public space is restricted within temporal boundaries (Sunday) and is set apart by social distance. They are closed off to Singaporean publics in defensive clusters, but at the same time segregated by social status. The behaviour of these Sunday enclaves of foreign workers, both on Orchard Road and in Little India, is regulated by physical barriers defining out-of-bound zones, as well as the watchful eyes of those who have jurisdiction over the space they occupy—be they the police or building managers. The imposition of control over these groups in public, including the constant surveillance of their behavior, put these Sunday enclaves within the mode of spaces that are restrained within the status quo.

4.2.2 Transferred Space

Transferred space is defined here as space in which practices are transferred in location, but without a change in form. These practices adapt to the new spaces but retaining much of the same social content. The spatial boundaries of these practices change, but not necessarily the temporal boundaries. Transferred spaces include spaces in which persistent practices from a cultural or spatial past are transferred into a new spatial environment which allows for these practices to persist with or without some form of adaptation to the new space. Transferred spaces thus include many of the public spaces of public housing where recurring everyday life routines inscribe spatial practices onto space which had been provided as a functional and largely neutral setting to high-rise living. The extensions of private property and activities into common areas like corridors and void decks of public housing, especially for the pioneering generation of flat dwellers, create spatial practices that become accepted norms within public housing. These practices may differ among the races and continue to change in the more recent new town developments, due to changes in demography and in the spatial form of the precincts and other new town spaces.

To some extent, the element of nostalgia operates in these transferred spaces, especially in the form of re-created practices and experiences, such as the experience offered to the tourist by taking a *trishaw* ride through the conserved area of Little India, following the prescribed routes and taking in the sights and smells of Indian culture within the carefully constructed experiential "-scape". The top-down re-creation of a traditional *wayang* stage in Chinatown,[3] as proposed by the Singapore Tourist Board, was met with criticism in the attempt to thematize culture. The 'naturalistic' *wayang* stages, set up by performing troupes often sponsored by community associations, are now found instead in the new towns to mark special occasions on the Chinese calendar.

Transferred space may also exist in the sense of free spatial practice existing within controlled realms of public space—it retains its location and meaning as a way of resisting change. This would apply to, for example, the concealed and contained red-light areas of both Orchard Road and Little India, where, through forms of spatial practice within demarcated boundaries, these realms of 'public spaces' resist change and persist in the extant forms of spatial practice.

4.2.3 Transgressed Space

In transgressed space, there is a temporary interrogation of established rules and accepted spatial practices. 'Transgressive' spatial practices may have the effect of a catalyst for change over time and repeated acts of transgression (Stallybrass and Allen 1986). The transgressions may take a 'polite' form—a de Certeau-sque tactic of the weak—that may have transformative effects over time. From the case studies, we encounter several of these transgressed spaces and practices in public. But more often than not, these only come across as glimpses or momentary lapses in practice, instances that bubble to the surface unexpectedly, and may or may not leave traces of its existence.

Transgressed spaces can be found in public space with dense networks of overlapping fields, such as in congested spaces. Transgressive spatial practices here involve stolen, borrowed or found spaces, where boundaries of private and public spaces may be breached, or where spaces are claimed and contested by different entities in space for their practices. These may exist in boundary conditions, as thresholds or interstitial spaces, where the needs of one type of practice overlap

[3]*Wayang* is the traditional form of outdoor opera performed by Chinese troupes on a temporary wooden stage set out on streets and open spaces.

with or impinge upon those of another. These conditions of negotiation or contestation of space often create sites of tension or friction due to forced intimacy. As these spaces tend to be tight and have concentrated activities, they tend also to be intensely used social spaces, where people are brought into close proximities, increasing chances of friction or social contact.

Another instance of transgressed space is witnessed through the phenomena of the Sunday enclaves, the temporary public space of transients in the city. The temporally-bounded appropriation by foreign worker groups of every available open space, including spaces not normally regarded as public space such as the back alleys of shophouses, imposes new forms of spatial practice within existing spaces, thus engendering a whole new topography of social spaces. These transformed public spaces witnessed at both Orchard Road and Little India demonstrate transgressed space as the manifest sites of spatial improvisation. The displaced spatial practices of these transient entities in public space tend to be outside of established or 'correct' behavior in Singaporean public space. The transgressions in space and in spatial practice bring to the foreground their deviation from the norm, rendering the norms more obvious. New meanings and spatial practices are borne out of the intense appropriation of space, which yield lingering influences and traces within the public space outside of its temporal boundaries.

Transgressed space exists as a result of defiant spatial practices in public housing. These may be instances of ignorance, creativity, resistance, or destructive and criminal behavior. The transgressed space exists in the temporal sense, for example, in the case of vandalism, or children ignoring rules regarding playing games in particular places; or as an experiential realm, for example, when a group of teenagers indulge in *le parkour*, or 'free-running.' "-Scapes" like those on Orchard Road may occasionally transgress norms of spatial practice, such as the playing out of flash mobs or the establishment of group identity by skate boarding through crowded zones of the Orchard Pedestrian Mall. These acts of transgression are temporal and serve to distinguish a particular group within public space through highly visible or even annoying spatial practice. Acts of vandalism or the throwing of killer litter represent destructive and undesirable aspects of transgressions in public space. These acts may be performed on the sly, so that the transgressor, although leaving a mark in public, is never seen or remains anonymous.

Acts of transgression on a recurring basis define a realm of spatial practice that may signal emerging sociabilities in space or highlight the presence of "-scapes" within public space. The often harmless, 'polite' acts of resistance, defiance or expression of difference in public space may lead to transformations in public space, or serve to bring to the surface subaltern groups and spatial practices. Transgressed space may not always have political content, although some of these spaces may be considered 'critical' spatial practice. An example of this would be the large group of nursing mothers who decided to breastfeed their babies in the public lobby of the Esplanade to protest against the management's unenlightened attitude towards the needs of breastfeeding mothers as mentioned in Chap. 2.

4.2.4 Transformed Space (or Space in Translation)

Transformed space occurs where there is a change in the form of spatial practice, with or without changes in the social content of the practice, although new social content may be introduced to the space. The spatial and temporal boundaries of these practices are transformed such that they may even be regarded as new practices. Altogether, sites, temporal zones and bodies in space may be supplanted within these transformed spaces. However, there are more instances of spaces in the process of transformation than actual transformed space. I will term these spaces in the process of transformation as 'spaces in translation.'

An example of 'space in translation' would be an Augéan non-place that has become a distinguishable and recognized public space. The underpass on Orchard Road leading across from the Scotts intersection to the underground Orchard MRT is such a case. The space has become defined by the same, visually handicapped busker and his accordion, belting out an almost predictable medley of Chinese songs, such that his absence is actually felt as some 'body' missing in the space. The space just around the underground station is also a favorite rendezvous spot for many young people—each feature in the space becomes a distinct place. The potentials of these transitory spaces are increasingly being recognized, as seen in the design of City Link (discussed in Chap. 2), which is considered a 'happening' space in the city and regarded as more than just another transitory device.

The temporal zones defined by Sunday enclaves temporarily transform public space through their distinct spatial practices and modes of congregation. In enacting new spatial practices of dense congregation, these Sunday enclaves have also elicited new approaches to negotiation on the part of the authorities. Instead of a ban on such congregations in public space, the phenomena of these enclaves have espoused new creative strategies by the authorities in the dispersal of crowds through the introduction of new activity grounds; negotiations between the needs of different groups; ways of minimizing and anticipating conflicts, and mediation of differences without attempting to eliminate these differences or to ignore them. These enclaves are also sites that demonstrate instances of the growth of civil society in the form of groups speaking up for the foreign workers' welfare and/or providing schemes to support and relief their needs. These public spaces thus become sites of friction and mediation between groups who are capable of exercising power to shape public spaces.

Some of the public spaces in the city operate in a sort of edge condition, in which different groups are brought into close proximity in space, in an ethnoscape. These are intense instances of public space functioning in different capacities for the different groups regarding them—as social resource, as places of identity formation, as information-rich references, for companionship and entertainment—in all, as spaces of exchange. These kinds of public spaces enable instances of 'friction,' when the ideas of self and of others are constantly evolving. The value of ethnoscapes as public space lies not so much in these entities 'meeting' in space, but rather in the dynamics of sharing public spaces offered by urban proximity. There is transformative potential in such sharing of public space as groups encounter the appearance of others whose spatial practices and views differ from their own. These encounters allow for shifts in perspective, regardless whether they result in exchange or antagonism.

Likewise, the flexible programming and the variety of public spaces found on Orchard Road bear transformative potential, as they enable individuals and groups to create their own itinerary of public spaces as well as the possibility of unexpected social encounters. No particular group is dominant in public space (other than on a temporal basis), and this condition enables choice and variety in public space. Together with a flux of social mixture and possible interactions, these features of public space allow for overlapping itineraries to inhabit the same location, and thus engender constantly evolving spatial practices.

The dis-embedding of images of elsewhere and the re-embedding of these within Singapore public space, as in the case of the hyperreal Christmas landscape on Orchard Road, is an instance of physical public space being transformed and in so doing, allows for the transformation of spatial practice. The embedded practices within public space are temporarily suspended, replaced with new perceptions of space, novel experiences and the heightened sense of 'event.' New nodes of activity emerge, subsuming existing boundaries of "-scapes," overlaying these with a new collective public space sustained over the festive season. Although the trappings and decoration of the simulated Christmas Wonderland is a case of temporary smoke and mirrors, the sense of wonder and transformation of space is real.

In the public spaces of the new towns, which have few precedents in the existing urbanism of the city-state, transformations of public space are created through adaptations of spatial practices to these new environments such that these practices take a different form. However, this does not mean that these practices have different social meanings, although entirely new practices have occurred in new configurations of spaces—how people move through them, interact within and inhabit them. New social places such as void decks, neighborhood coffeeshops, food centers, markets, and transit hubs are nodes within channels of activities and constitute a realm of public spaces in which spatial practices are re-enacted on an everyday basis—ultimately becoming embedded in space. The configurations of space in the new towns create a spatial system that is rather dispersed, but converge at the town centers and in the channels of movement within these centers, concentrating people within these orchestrated routes and nodes. By mapping the daily routines of new town dwellers, one identifies well-populated 'temporal zones' or 'public time,' when there is the greatest likelihood of meeting neighbors in public spaces. The reconstituting of public spaces within the rhythmic flux of public time is a case of transformed space in the new towns.

New connections in relation to public spaces such as that afforded by the mass rapid transit system bring people from one set of public space network to another within minutes, creating an almost seamless transition from one space to another. An example would be the transition from the new town public space of Toa Payoh to the public spaces of Orchard Road within the temporal space of about twelve minutes after entering and exiting from the portals of the MRT stations. Such a conflation of domains of public space, especially of the city and the Singaporean heartland, transforms perceptions, accessibility and actual space due to the exchange of entities from one space to another.

4.2.4.1 Public Space Dynamics

By plotting on a graph the trajectories of "change" versus "control," four modes of public space, represented at the extreme polarities by tradition, transfer, transgress and transform, are established. Within these modal frames, the previously discussed themes in public space are inserted to situate these within the differentials offered by the axes.

The graphical representation of such a mapping shows that there are more themes uncovered from the case studies within the right-hand side of the map. This suggests that there is a trend towards change, whether these are transformative or transgressive in nature. Read diagonally from the left bottom frame towards the right upper frame, the mapping shows the bulk of themes falling within the frames of the transformative and the traditional, with a trajectory towards the top right of the map, i.e. towards transformation. The frames for transfers and transgression may be read as spin-offs or 'ricochets' from the main trajectory of the traditional towards transformative (or its 'in-process' form, *translated* space). These can be read as modes of 'coping' with change, through transferring and transgressing, rather than as final states of equilibrium in spatial dynamics.

Although the dimension of time is not present in the mapping of the framework, the general historical trajectories reflected in the development of Singapore public space root for an overall movement towards translation and ultimately, transformation. The mapping captures an intermediate phase, or even an equilibrium position within such a constant flux. In other words, it is unlikely that the map would show a total migration of themes into the modal frame of transformed spaces, as there will always be some form of resistance to change in the form of control by those who have power over space and spatial practices. As such, there will also continue to be coping mechanisms such as transference and transgressions to serve as intermediate modes of spatial dynamics. The graph illustrates a constant state of flux that is stable within long spans of time, as public space itself is the site of negotiation of change, and spatial practices are embedded within the structures of these spaces.

It would be incorrect to assume that the government is always seeking total control of public space and is resistant to change. Conversely, it is also untrue that Singapore people are always for change in public space, and always seek its transformation. At this point, it will be unhelpful to assume that the government and the people are in polar opposition and take antagonistic positions with regard to the dynamics illustrated by the mapping. Instead, the dynamics of Singapore public space are fuzzier and more complicated than such an essentialist argument could explain.

4.3 Situating Singapore Public Space—Bases of Difference

With a clearer picture of the forms of public space, spatial practices and their modes of operations, it is pertinent to relate these back to the larger social, political and cultural milieu of Singapore. In situating Singapore public space, the question that most comes to mind is: what are the uses and functions of public space? What are the modes of transmission and mechanisms through which public space operates in Singapore? Finally, what distinguishes Singapore public space from classical rational-critical definitions of public space?

4.3.1 The Uses and Functions of Public Space and Their Modes of Transmission

4.3.1.1 National Identity

One of the functions of Singaporean public space is to forge a national identity. While the state constructs the dominant rhetoric of the Singapore identity, the very constructed-ness of this identity makes it subject to negotiation and contestation by different groups or individuals (Kong 2003, p. 2). The site of these negotiations is often public space. Instead of overt confrontational events in public, these negotiations often take place through ordinary public life, through the bodies that inhabit public space and in their recurring practices. Notions of "nation" and "identity" are constructed through spatial practices by the state, and through people's lived experiences in Singapore public space.

It is particularly in the quotidian landscapes such as those of public housing that the concept of public space is taken for granted and is thus ideal for "ideological appropriation" by the state (ibid., p. 3). The notion of the Singaporean heartland and the Singaporean Dream are constructed within these spaces, including those spaces that remain defiant and resistant to such notions.[4] The understanding by the Singapore people of public housing space as the "exclusionary space of national narratives," and the state's use of public housing in the material space for nurturing ideal Singaporean "communities" already make these spaces heavily invested with function (Yeo 2004, p. 27). However, it is clear that a close reading of public housing spaces goes beyond interpreting these spaces as the reification of power relations between the state and the people. These spaces allow for the insertions of negotiation through forms of resistance, collusion, translations of meanings and transformations within the sphere of everyday life.

[4]To have the five Cs—condo, club, car, credit card and career. These are the explicit material gains that a Singaporean could possibly wear as the badges of success in a society often defined by "conspicuous consumption".

The formation of a national identity also involves the formation of a palpable sense of being "one people". However, Singapore society is a "low-trust" society as the low levels of "social capital" of the culturally disparate groups make it difficult to develop natural wellsprings of trust and cooperation.[5] The problems of a multiracial, multicultural society become apparent in public space, especially in the context of high density living in public housing. Public space shows up the negative aspects of Singapore society, especially in the lack of trust between fellow citizens, and the reliance on the state as the arbiter of socially accepted behavior.

The widely held notion of the Singapore Dream, embodied within the notion and physical space of public housing, is really about fleeing from a life in public housing, aspiring to own private property, playing in the private playgrounds of exclusive clubs, traveling using one's own transport over public transportation, and achieving status and means beyond one's fellow men. The role of public space in public housing offers an alternative view of the Singapore Dream: one in which the community, consensus and conflict may play a part in the constitution of care and (multi)-culturalism, rather than in purely personal material gains. It is interesting to note the almost parallel development of the precinct community space in public housing and the emphasis on Shared Values as instituted by the state in the 1980s. The notion of 'Asian' was associated with communitarian values and the notion of 'Western' with the tropes of materialism and self-interest.

So what constitutes the Singapore identity? It is not always easy to construct a representative type for this notion, which is manifested in a multitude of forms in the individual lives of Singaporeans. What may be more readily construed would be the government's vision of what such an ideal construct would be. This seems to embody traits that may at the same time be contradictory if they exist in an individual: tough-minded, modern, cultivated, cosmopolitan, enterprising, competitive, yet able to relate to 'Asian roots' and 'traditional values,' communitarian and family-oriented. The schism implied from this set of characteristics may be inferred also from the lack of a middle ground in terms of the resolution of public space. The transition from the cosmopolitan and urbane spaces of exchange in the city to the socially-engineered, communitarian spaces of public housing is an abrupt change of spatial environment.

Individuals embodying all the ideal traits specified by the state are rare, and more often, we encounter in contemporary Singapore the "Cosmopolitans"—those who call the world home and are cultivated individuals who embrace their international orientation but may not always relate to their stay-at-home counterparts, the "Heartlanders". "Heartlanders" include the taxi driver or hawker who provides essential services in the local economy; the benign "aunties" and "uncles" whose favorite hang-outs are the neighborhood coffeeshops who tend to speak in dialects and are uncomfortable outside of their realms of daily existence[6]; the young

---

[5]The British colonial government divided and ruled by compartmentalizing the races and dialect groups.

[6]It is a peculiar Singapore custom to call older, non-related women and men aunties and uncles.

Singaporean "Ah-Bengs" and "Ah-Lians"[7] who grew up in the 'heartland' and favor western pop culture, fads and fashions, such as listening to hip-hop and sporting D&G tops and Evisu jeans, and speaking in "Sing-lish"—a local adaptation of the English language, with its own intonations, abbreviated grammar and words borrowed from Malay or the local Hokkien dialect; the conformist, conservative civil servant who smiles and provides friendly service because of government campaigns to encourage such behavior; the materialistic condominium dwelling Singaporean family, who works hard to pay the mortgage for their apartment and their expensive cars and have time for little else; and the people who dwell in the fringe of acceptable society, such as single mothers, former drug addicts, school drop-outs, pimps and prostitutes, and chronic failures who fall out of the system of meritocracy.[8]

4.3.1.2 Cultural Identity

Through the case studies, Singaporean public space has been established to be a potential repository of culture. Public space is the site of the re-enactment of spatial practices that continue to define and distinguish one set of cultural selves from others. These spaces embody complex dynamics and the overlay of spatial fields that define the subjects within space, establishing forms of commonality, particularity and difference. Public space also manifests gendered structures within cultures and the power relations between different groups in space. Concepts of 'we' and 'us' against 'them' and 'others' are constantly being defined and re-defined. The ability of public space to become embedded with particular forms of spatial practices, support and allow for recurring enactment and reproduction of cultural public spaces.

On a large scale, public space is often used as the manifest site of the cultural status quo through the public spectacles of processions, festive decorations and events. These spectacles are often state-sponsored and serve to remind Singaporeans of their Asian roots and also the notion of a multicultural society. The staging of festive events and celebrations in different quarters of the city traditionally associated with different cultural groups establish spatial fields distinguishing different groups, and define Orchard Road as the portal through which 'foreign,' mostly western, cultures set foot in Singapore public space.

The identities of different groups within the city are established through spatial practices in public space. Some groups may appropriate particular spaces or establish spatial practices that distinguish them from others; other groups may meld into public space and exist as entities within the ethnoscape. Sometimes, the 'being

[7]These are terms, with a slight derogatory tone, referring to the not-so-sophisticated tastes of these youths and their fake-blond girlfriends, who are often togged out in fashion paraphernalia, and who drive souped-up "Beng-mobiles", with tail fins and bright, flashing blue-lights on the rear.

[8]In other words, the types of individuals found in Singapore may be found in other parts of the world, in other cultures, called by other names.

apart' of groups in public space comes not so much out of choice but as a manifestation of social distance or involuntary 'enclaving.' However, due to the "culture of congestion" that exists in Singapore public space, groups may share boundaries or have to negotiate the sharing of space (Koolhaas 1978).[9] The cultural identities of marginalized and dis–embedded groups in the city, such as the low-wage foreign workers from South Asia, Thailand, Burma and the Philippines are re-embedded in public space, through association with particular spaces in the city and through recurring spatial practices in these spaces. Through the recreation of their own spatial practices in the Singaporean context and their claims to space, on which their physical existence and right to the city is staked, they have claimed representation in the city—a space from which to 'negotiate.'

The multiracialism is best depicted in public space not so much in the interaction of the races, but more so in the sharing and negotiating of the use of 'play' spaces, especially in public housing, where there is an absence of dominance by any one group. Clear principles such as first-come–first-served, sharing space for simultaneous play, joining in, or agreeing on the duration of the day to use the space apply (Lai 1995, pp. 110–111). Although the public spaces in the city associated with the different cultural groups are found apart from each other, it is not uncommon to see a mingling of races within the cultural districts. An interesting example of cultural multiplicity can be seen in the Rochor area, where a Chinese and an Indian temple are in close proximity on the same street, and visitors of the two races often offer prayers at both, perhaps to increase their chances of good luck and prosperity. As we have also seen in the Little India public space, some activities are not defined by race but by gender and inclinations.

4.3.1.3 Individual Identities

Individual identities are established in public space through the experience of space for oneself and the drawing out of personal narratives through urban experience. This can take the form of assembling a personal spatial itinerary or space *a-la-carte*, thus expressing personal preferences through the use of space, or in viewing the urban through a particular form of spatial experience, such as *parkour*—taking urban space as a challenge by moving through using one's athletic and manipulative skills. The documentation of personal experiences in the form of Internet blogs has become an important phenomenon in the narration and account of public space, with these providing an infinitely rich-source of information about individuals' perceptions in space. Experiences of individuals in space are not limited to the consumption of space in leisure activities, but also through working within public spaces, with these individuals themselves becoming a feature of public spaces.

---

[9]The term 'culture of congestion' was used by Rem Koolhaas in *Delirious New York*, 1978, to describe the new urban culture of the contemporary metropolis, one in which innovative ways have to be invented to circumvent the shortage of space.

4.3.1.4 Political Legitimacy and Control

Public space, through the practice of everyday life and allowing for the multifarious readings of spatiality, gives insights to the constructions of the notion of ‘nation’ in more concrete terms than through the abstracted notions embodied in the dominant discourse offered by the state. Public space is also the site of potential conflict and negotiation, resisting or at least making malleable embedded structures of boundedness and control within space, embodied in its physical form or content.

One of the functions of public space from the perspective of the state is in the imaging of abstract notions of “Tropical Garden City,” “gracious society,” “community” “world class city,” “global city,” “Renaissance city,” among other incarnations of the desired form of the island-state, depending on the state of development of its history. Public space enables the material representation of these notions, mainly through aesthetic means as the parameters within these can be controlled. An example of such a public space would be the Padang, which over Singapore’s history of developments and especially in its ‘founding moments,’ has played a crucial role through its imaging. In different moments, it had framed notions of colonial sovereignty, independence, triumph, nationhood, solidarity, confrontation, culture, and recreation.

The strategic location of the Padang as a green, open space in the city provides the ideal foreground to the Central Business District, framing the ideal poster for the coming of age of the global city, a symbol of economic progress. Orchard Road has formed the backdrop for the city as a ‘world-class’ location for shopping, of urban culture, and of the notion of ‘happening.’ The public spaces of the ethnic quarters have provided the setting for a city concerned with its heritage and multiculturalism, a city embodying a ‘gracious society.’ The public space of public housing provides the imaging of Singaporean communities in the ‘heartland,’ the basis of depictions of the multiracial society and the idealized landscape of “the best home,”[10] with a “world-class” standard of living.

While the imaging of public space is well within the control of the state, spatial practices are less easily bounded to fit into the paradigms drawn up by the state. While boundaries of acceptable practices are drawn in public space either in the form of concrete barriers or in the intangible forms of rules and regulations, implied social consensus is also embedded in physical space as infused ideology. Power relations are played out in public space, among the groups that occupy public space, as well as between the state and its people, through the interpretations of the dominant discourse and its inflections that are enacted through spatial practice. The negotiations in space do not always assume that the state is dominant and its people are dominated, or that marginalized groups in public space are subservient to mainstream Singaporeans—these positions shift according to how the situation is framed. While strict laws regarding the rights to assembly in public space still apply

---

[10]The phrase, “make Singapore your best home” was coined by Prime Minister Goh in the early 1990s.

and the Speakers' Corner is more symbolic than functional, the question we should be asking is not so much if democracy is missing in Singapore public space, but the form it takes, and the kinds of urban and spatial dynamics through which these processes are played out.

4.3.1.5 Public Space as Source of 'Knowledge'

Public space, just as it can be a source of imagery for the state in the state's crafting of the ideological landscape, also acts as a repository of information or 'knowledge.' Unlike other consumable commodities, 'knowledge' is a public good that can be shared and exchanged without impoverishing its contents. It would be useful here to consider this function of public space in relation to the different kinds of knowledge that it beholds. These forms of knowledge can be thought of as *know-what, know-why, know-who, know-when* and *know-how.*

Public space establishes a realm of common knowledge or 'common sense' about Singapore society, governance, culture and general sensibilities—the *know-what*—as its operative milieu. We have already found that public space can act as a repository of culture at many levels, and resonates both physically and structurally with the forms of representation of the state's attempt to direct the national culture. The *know-why* component problematizes this aspect, in the interpretation of the outcome of such an operation. For example, the lack of public dissent and protests in public space can be attributed to the operative milieu of the style of government, the form of Singapore society and the history of the development of public space.

The *know-where* component of public space is that which defines 'place' through spatial categories such as "Singapore," "the city," "Chinatown," "new town precinct," "here," "there," etc. The *know-who* component deals not only with the information about other people and groups in public space, but also with establishing one's own identity in public space. Identity is defined with reference to one's local knowledge and spatial practices, such as if one is an "insider" or "outsider" in a particular space. The *know-when* embodies the temporal elements of public space, defining the 'then' from the 'now' and the long and short-term cycles, as well as the rhythms played out in public space. It also foregrounds change and transformation, with the present as the frame of reference.

The last component, *know-how*, is two-fold: first, it represents how all this information is managed and communicated within public space; second, it represents how space is produced and re-produced, i.e. the existing dynamics within public space and spatial practices which maintain these within its operative milieu. In other words, *know-how* represents the mode of transmission of information embodied in public space, and the mechanisms through which these public spaces function.

4.3.2 Mechanisms: Public-Space-in-Practice

The method of this study has been to construct notions of public space through spatial practices, i.e. actions of bodies in space, enacted on a recurring basis. Instead of taking physical public spaces as a starting point, this study asks if embedded structures in public space are played out through activities in space. Spatial practices become the means through which the emergent structures of public space are constructed in recurrent spatial interactions. Human actions in public space can either play out according to the way space has been designed, or can change design intentions for the space through operations of transferring, transgressing or transforming. The emphasis on spatial practice put bodies in space as the transformative subjects who enact change through their actions and engagement with space. The recurrent spatial practice shapes the embedded structures of public space, such that these practices 'produce' the space just as they are shaped, i.e. facilitated or constrained by it. The emergent spatial practices enacted within public space on a recurrent basis can also be described as *public-space-in-practice*.[11]

Because *public-space-in-practice* is often enacted within overlapping social systems extant in space, spatial practice often embodies invisible social or cultural structures. For example, in the gendered spatial practice enacted in Little India, the gender-biased social structure embedded within public space emerges. The dominant discourse of multiracialism is also enacted in space when people of an "outside" ethnic group claim their rights to public space, such as the Chinese residents of Little India reinstating their rights to space claimed by the 'dominant' Indian groups.[12] The history of public space development in Singapore has shown the state's urgency in dismantling spaces of dissent (i.e. those spaces which allow the enactment of the spatial practices of racial communality, such as *kampongs*), as well as the state's agoraphobia in producing new spaces which have potential as large assembly spaces, especially in the city and on campuses, to avoid public rallies and protests. The state's dismantling involved not just of physical space itself, but also of social structures, which enable and empowered groups based on racial or religious inclinations.

Spatial practices may, as described earlier, reinforce the status quo, in acknowledging the dominant social structures of these spaces. Or they may transgress or circumvent these structures, or even transform them. For example, by changing how barriers are perceived in new town spaces, *parkour* practitioners transgress the norms of moving through new town spaces, thus creating their own spatial itinerary in negotiating these constraints. Also, old people ignoring the rules evoking a satellite living room through bits of furnishing and seating create a void deck culture in new town precincts. They appropriate public space and transform

[11]To come up with this term, I saw the parallel development described by Wanda Orlikowski in "Using Technology and Constituting Structures: A Practice Lens for Studying Technology in Organizations" (2000).

[12]Here, this means the majority.

these into new social spaces. In all these instances, change is evoked by human actions in space. The actions of old people in void decks may eventually lead to more transformations, such as the Town Councils perceiving the need for places for seniors to socialize, and the need to provide these places in the newly constituted void decks as designated "Senior Citizens' Corners," with built-in seating and tables.

Changes in spatial practices may be brought about by need, improvisation or opportunity. For example, the gathering spaces on Orchard Road have become congested, so some foreign maids have found new gathering spaces in the new town centers, which offer more space and also more affordable food and commodities, leading to an emergent new pattern of spatial practice in the town center. The phenomena of Sunday enclaves in both Orchard Road and in Little India are instances of improvisation. Walkways and road curbs become seating areas for the Filipinas' rendezvous, while the edges of drains form makeshift benches for south Indian workers. While railings mean barricades to walkers, *parkour* perpetrators see these as opportunities to show off their skills of spatial imagination.

A unique case, albeit subversive, involving need, improvisation and opportunity, is witnessed in the new behavior of illegal gambling in almost public locations. As gambling is illegal in Singapore, operators of gambling dens have constantly been busted by the police while conducting their activities in fixed locations. On the other hand, at Chinese funeral wakes, often held in the void decks of housing estates, it is not uncommon for the relatives of the bereaved to play cards to pass the time during these wakes, which may last for days. Den operators have seized the opportunity to set up improvised gambling dens under the cover of a real funeral wake, as police usually do not arrest family members who play cards at wakes. In this sense, the new 'spatial practice' in public space offered a newfound mobility to these furtive operators in their inserting these activities within existing accepted social structures (Straits Times 2004).

It is essentializing to read spatial practices in public space as a form of power relations between the state as the dominant power, and its people as collective, manipulated subjects. Such a reading does not allow for the observed phenomena of collusion, conflict, collision, counter-strategies and negotiation that occur in the realm of everyday life (Kong 2003, p. 11). While direct confrontations in contemporary Singapore do occur, they are rare. More often than not, they result in heavy-handed efforts on the part of the state to clamp down on these instances of perceived dissent and chaos.[13] It is more likely that conflict and negotiation occur at the level of the recurring practice of everyday life, and through *public-space-in-practice*, eventually leading to some form of transformation in the public sphere. It is by its insertion within everyday life that *public-space-in-practice* plays the role as a potent agent of change.

---

[13]This was witnessed in the general election of 2001, where political protesters, who campaigned on the streets and blocked traffic were arrested on the spot. Public protest is still illegal in Singapore.

4.3.3 Bases of Difference: Singapore Public Space

The highly centralized hierarchy of government in Singapore, with its top-down decision-making approach with regards to policy making and national development, had over time, not been conducive to the growth of an independent civil society which mediates between the public life of citizens and the private sphere of the family (Ooi et al. 2004, p. 80). The state is aware of the need for civic engagement, as the detachment of Singapore's citizenry from the decision-making processes made in relation to the development of the country would mean that its people would have less of a stake in ensuring the long-term success of the nation beyond their personal gains. Continued disenfranchisement of its people would espouse an attitude of indifference towards the nation, treating as a hotel, from which one could come and go as one pleased (Yeo 1991, p. 81).[14] Forms of civic engagement have been carefully contained within government-linked or sanctioned organizations (Ooi et al. 2004, p. 76),[15] dealing with spheres of influences including consumer rights, to environmental action, to ethnic self-help groups (ibid., p. 75).

The depoliticized Singapore society has also been cast as a victim of its own success—the state's efficiency in meeting its citizens' basic needs such as housing, education, health-care and transport, did not create a climate leading to the empowerment of its citizens to act politically (ibid., p. 75). Singapore's denizens have been criticized as too comfortable and materially well-endowed to mobilize politically, even as they complain incessantly of the ubiquitous, paternalistic state, as described in the article "The curse of being Singaporean":

> This is the curse of being Singaporean…But wait, you ask. What's so cursed about it? Well, the limits of living in an island so tiny your soul finds it hard to breathe sometimes. The limits of living in a society which defines success primarily by material attainments. Most of all, you suffer the pangs and angst created by the presence of an all-powerful, dominant state (Straits Times 2004).

Due to the growing social and cultural diversity of Singapore society today as Singapore moves into the global labor market and establishes itself as a 'world class city' for business and for the arts, the government of Singapore is hard put to remain centralized and rigid in its forms of governance and decision-making processes. The nature of information distribution and exchange proliferates beyond the censorship and control of the so-called "nanny-state". It is becoming more likely that the structures of power will move towards a more diffused state.

---

[14]Minister George Yeo referred to the need for civic society, or else Singapore would remain nothing more than a well-run hotel.

[15]The state's view on civil society organizations eschews activism and limits civic participation to forms of welfare voluntarism. The state is ambiguous in its support for independent and non-government led initiatives.

4.3.3.1 Public Activism—The Cases of Chek Jawa and the Rail Corridor

Public space remains the important site of negotiation between the state and its people. Although many lament the lack of active civil society in Singapore, public space stands as the mute but active agent of social transformation. Instead of acts of civil protests and confrontation in public space, the mechanism through which transformation takes place is recurrent and persistent practices enacted in spaces situated within quotidian landscapes, instead of spectacular landscapes such as the Tiananmen Square in Beijing. In the aftermath of the 9/11 events in the United States and the decision of the US to go to war with Iraq, popular demonstrations had filled the streets of major cities in the world. In Singapore, a few would-be protestors were stopped by police near the Tanglin Shopping Mall, en route to the American Embassy, with their placards still in plastic bags (Straits Times 2003)—"(t)he globalized world of political protest encountered a blank spot in the city-state as the PAP government gave its tacit support to war" (Ooi et al. 2004, p. 142).

This is not to say that outright activism does not exist in public space. Public activism, or what we might term "critical practice," was evoked in the case for the conservation of Tanjung Chek Jawa, one of the precious few remaining marine habitats in the coastal areas of Singapore (Ooi et al. 2004, pp. 133–136). The Rail Corridor is another site where 'critical practice' has more recently contributed to the preservation of the former Singapore-Malaysian railway line as public space. These two cases can be regarded as successful instances of public activism in Singapore, demonstrating how specific engagements between civil society and the government authorities pertaining to public space can lead to transformative developments.

The significance of Chek Jawa case was that "(h)ailed as an exercise in constructive engagement with government authorities, the nature of public activism in this case is likely to be used as a blueprint for popular protest in years to come" (ibid., p. 133). The Chek Jawa example had even been optimistically cast as the "turning-point in the relationship between state and civil society in Singapore," as this was a case in which government policies were changed due to public activism (ibid.).

In the 1992 Masterplan of Singapore, the government had planned to reclaim land in the eastern part of Singapore for military training purposes, which would have destroyed the extensive sandbar of Chek Jawa, just off Pulau Ubin, one of the offshore islands of Singapore. Local nature lovers, aghast at the potential loss of the natural habitat of unique marine plants and animals, harnessed the efforts of specialists, enthusiasts and volunteers to organize expeditions, give presentations, and mount exhibitions. They generally promoted the Chek Jawa cause to both the decision makers as well as to the public.

The group, calling themselves "Friends of Chek Jawa," carefully avoided the term "protest" and mounted a low profile and sustained campaign, using catchphrases like "Remember Chek Jawa" and "Chek Jawa is Alive!" instead of "Stop the Landfill!" or "Mindef (Ministry of Defense) Keep Out!" which would have jeopardized the friendly tone of the campaign. They used terms like "nature-lovers" and presented images of families with children enjoying the sights of starfish, sponges and sand-flats, emphasizing the educational value of retaining such a place in urbanized Singapore.

The convivial tone of the campaigners were more akin to that of a Sunday picnic than of people chaining themselves to rocks or demonstrating violent forms of protest. The "Friends" generally avoided the use of adversarial terms or the use of a confrontational approach. Instead they took on the demeanor of civil debate backed by information, emphasizing the potential public good that could be derived from the conservation of the site.

Their efforts were rewarded with the deferment of the reclamation, originally scheduled for the end of 2001, "for as long as the island is not required for development".[16] The form of spatial practice used by the "Friends" turned the site of Chek Jawa into convivial public space. The soft-sell approach, involving not just the nature lovers, but school children and families on a Sunday visit, evoked a kind of *public-space-in-practice* of everyday life, which led to transformative results. The campaign would not have worked if a different approach were adopted. Although other reasons such as the nature of the case and its location far away from scarce city sites would have played a role, the case of Chek Jawa illustrates the kind of public activism, which approximates the negotiation made with the state by emphasizing spatial practice enacted on a recurrent basis, which succeeds in transforming space. In this case, this meant to leave Chek Jawa as it is.

Formerly the Keretapi Tanah Melayu Berhad (KTMB) railway line connecting Singapore and Malaysia, the Rail Corridor is a 26 km long 'corridor' that was left relatively untouched after rail operations ended in 2010. The Nature Society proposed to preserve the corridor as a 'green' space, eliciting support from bloggers and netizens who utilized websites and other forms of social media, such as Facebook, to publicize the cause. The Ministry of National Development (MND) and the Urban Redevelopment Authority (URA) also recognized the call to preserve the green corridor as an opportunity to engage the public for their ideas in redeveloping the corridor.

Both civil society groups and the MND and URA have organized events and initiatives that invited the public to contribute their ideas towards re-envisioning the Rail Corridor. These events and initiatives included the setting up of an advisory group, the Rail Corridor Consultation Group (RCCG) that is chaired by the Minister of State (National Development), Tan Chuan-Jin, and includes members from NGOS, cyclists and nature bloggers; the "Reimagining the Rail Corridor" exhibition by Friends of the Rail Corridor; and URA's 2011 ideas competition "Journey of Possibilities". These developments shaping the rail corridor as a public community space are still ongoing. Today, the Rail Corridor stands as a testament to the positive state-public engagement that led to its preservation (Fig. 4.2).

---

[16]This phrase is cited from Ministry of National Development statement on 14th January 2002, as reported in "Reclamation works at Pulau Ubin deferred," *Skyline* January/Feburary 2002. https://www.ura.gov.sg/skyline/skyline02/skyline02-01/text/skylinePBack.html.

Fig. 4.2 Panorama of Rail Corridor. *Source* Author's collection

Some commentators have argued that "(t)he strategy that has been developed in managing civic life in Singapore has been, in large part, a spatial strategy" (Ooi et al. 2004, p. 80); and others have commented of Singapore that, "(i)t is not so much that democracy is missing in Singapore but more that the question of democracy has been made redundant—and this is a problem that urbanism must face: the intensification of urban processes, which once certainly thrived on the positing of democratic ideals, increasingly renders democracy irrelevant" (Bishop 2004, p. 7). While one argument posits that space embodies some form of civic ideology, the other claims that ideology may no longer be embodied in space, which has taken its own trajectory.

4.3.3.2 Change and Public-Space-in-Practice

My argument is that while space might have been effective in facilitating or constraining forms of civic representation—such as large, centrally-located spaces being often the sites of protests in the city—the phenomena of *public-space-in-practice* has developed its own trajectories, such that its enactment is potentially space-shaping instead of being entirely controlled by spatial form. While public spaces are often the sites of spectacular civic representation, the more powerful transformative agents are the *public-space-in-practice* situated within everyday life, which produces and changes public space. Spatial practice and public space are supplemented by other forms of non-physical public realms such as the Internet forums and the Forum pages of the local papers, through which topics concerning Singapore society are discussed and debated.

The direction in which *public-space-in practice* develops shows the cartographic trajectory of change from traditional space towards transformed space (see Fig. 4.1). Physical urban space itself has been rapidly transformed, as urban development in Singapore often means demolishing the old and replacing it with new urban forms and programs which increases the economic value of the site. As such, urban planning and control had been instrumental in Singapore's growth as a development city-state. While the attitude of the state towards 'old' buildings in the early years of independence were that few were worthy of preservation (Choe 1969, pp. 165–166), the emergence of the discourse of "nostalgia" in the 1980s and 90s was seen as a response and resistance towards the fast pace of economic development, the rising materialism of Singapore society, and the "industrialization of everyday life" (Kong 2003, p. 132). City boosterism, the selling of Singapore as a venue for international conventions, exhibitions and tourism, made conservation of old buildings and districts economically viable, especially with the official tourism sales-pitch of 2004 being "Uniquely Singapore".

Singaporeans, at that moment in the growth of the nation, felt a sense of having arrived, but also felt that their city might be lacking in a certain spirit and soul (ibid.). The discourse on nostalgia, therefore, was a condition arising from its present state of development. It was more of an anxiety about the lost past, which a new generation of Singaporeans might not even remember, than a desire to live in a historic, museumized city.

However, the critique on relentless development did not mean that the state had lost its perspective on the constant renewal of the city; it was also the case of the state desiring a balance of continuity and change. The state is invested in the physical transformation of the landscape and is in fact, at the forefront of necessary changes to its structures, and the instruments of statehood, which would sustain the nation in a more competitive economy. With regard to national development, the state has also adopted more consultative policy-making processes, with the involvement of focus groups and public debate in the drawing up of the Concept Plan 2001. This is in contrast to the history of Singapore's development, in which the state had unilaterally changed the whole paradigm of the home environment in the nation in one fell stroke, when it demolished slum housing and built the first new town.

Religious groups and concerned individuals had also contributed actively to the recent debate about allowing a casino to be built in Singapore, with most of these groups lobbying against such a move, making this the most discussed topic of 2004 in the Forum pages of the Straits Times. The government had for years stood its conservative ground in not allowing such enterprises to take root in Singapore, but now seemed game to consider this as a boost to Singapore's position as an entertainment and leisure hub, and had tried to push the debate beyond 'money versus values' to whether Singaporeans can act responsibly. The lobbyists, however, remained persistent in their resistance to such a move, but realized that the form of resistance had to be carefully construed:

> 'In Singapore, making a lot of noise in the public arena doesn't go very far,' said President Tan Thuan Seng of the local offshoot of a United States-based movement that promotes Christianity by preserving traditional values. 'So we are doing it quietly, writing directly to them' (Straits Times 2004).

However, not everything is subject to change in the city-state: certain aspects of the political landscape of Singapore remain in the status quo, such as one-party rule, the limited freedom of the press, the constraints on the discussion of tabooed subjects of religion and race, and the power relations between the state and its people. Conversely, as shown by the debate over the building of the casino, Singaporeans are not always game for change, and are not always the agents pushing for change. The central tenets of the government, embodying good economic growth, tempered by multiculturalism, meritocracy and a 'gracious society' steeped in Asian values, indeed find many supporters among the Singapore people.

Chapter 5
Rethinking Public Space: The Singapore Model

The site of visible transformation in Singapore is often public space, although non-physical 'forums' of negotiation increasingly play important roles in articulating these transformations. Public space is also the site for the state to claim its legitimacy in the normalizing of social relations in these spaces of everyday life, so that the power relations with the state are 'naturalized' and assimilated in quotidian space, and the status quo is maintained. As the site of transgressions, public space defines the norms of spatial practice as well as 'publicizes' the inversion, breach or transformation of these norms.

5.1 Towards a Singapore Model of Public Space

The classical ideal of Habermasian[1] "public space"[2] is based on "rational-critical discourse", a condition only made possible with communication instead of domination. "Communication" in such a context means not only sharing ideas and information, but a whole process in which "reason is advanced by debate", leading to potentials of transformation (Habermas 1989). In such an ideal discursive realm, the public would act as a large-scale entity with converging aims, and consist of individuals interested in achieving change for the common good—assuming that these individuals are suitably represented and are articulate with their ideas and wants. Jürgen Habermas' idea of public space is concomitant with democratic political practice, and assumes the existence of an independently acting civil society, separate from the state and from the private realm. Based on such an ideal,

[1]I use Habermas' pre-transformation model as the classical model for discussion. In the discussion of Habermasian "public space", the abstract idea of space is evoked, more in the form of an ideal public sphere than actual space. It is also clear that Habermas describes the subsequent disintegration of the public sphere in the course of modernity.
[2]Here I use the term space to embody notional as well as the physical space of being.

L. Hee, *Constructing Singapore Public Space*, Advances in 21st Century Human Settlements, DOI 10.1007/978-981-10-2387-3_5

abstracted model of the public sphere, Singapore public space is clearly problematic, in that:

- The state produces the dominant discourse in public space, and the form of "Asian" democracy practiced in Singapore is not really congruent with the western definition of democracy.
- The form of multiracial public in Singapore society, with its vastly different and sometimes antagonistic interests and levels of representation, falls short of a singular public acting on the common good, although the Singaporean identity is beginning to emerge in the new public debates over the last five years.
- There is only a weak form of independent civil society in Singapore—most organized and funded groups are government initiated or government linked.
- "Negotiation" in Singapore public space is "soft"—there is only an inchoate discursive public space based on verbal debate.

The state remains the dominant entity in Singapore public space, exercising its influence through government and decision-making processes, and as the dominant producer of physical space. Although a more consultative approach has been adopted for some aspects of Singapore's development, the modernization and westernization of Singapore did not extend to the adoption of western liberal democracy and a culture of citizen participation. The trope of the Asian value system still rules: communitarianism and respect for tradition as the 'moral compass' against the so-called 'western' traits of individualism and materialism.

The state's legitimacy is deeply rooted in the notion of survival—Singapore's totally contrived existence in the world economy is cast as vulnerable to global forces, and its continued prosperity is founded on the existence of a strong, incorrupt government capable of making the best, if not always the most popular decisions, for the nation. The state often defends its moves by evoking the 'exception rule'—e.g. the notion that we have to be different from western democracies if we are to survive as a small nation without natural resources; or the notion that we cannot allow the free press to undermine the integrity of the government.

The rule by consensus, i.e. consensus of the people with the government, had been the basis of the Singapore government since its inception, with the domination of the state as the primary condition of the developmental state. The overarching risk adverseness, termed "*kiasu*-ism" in local lingo, color many aspects of Singaporean life, and already renders any negotiation with the government on an unequal basis. It is not without reason that Singaporeans in the past have been described as apathetic to many issues that would in other parts of the world, have generated heated debate or protest. This, however, has begun to change in recent years, with the new 'normal', and an evolving civic society evoking debates over many issues, such as protecting green heritage, place identities, becoming a car-lite society, etc.

Singapore society, as previously elaborated, can hardly be construed as a large-scale single entity, even if the argument of the mostly middle-class population is assumed. The multiracial society has diverse interests divided by race and religion, and exists within a social milieu of 'low-trust' and a general wariness, especially in the aftermath of 9/11, when elements of regional Islamic radical cell, Jemaah Islamiah, were arrested for a plot to sabotage Singapore. The acknowledgement of the lack of real racial integration and the failure of the racial distribution policy in public housing were foregrounded, when the government, in its post-9/11 advisory to citizens, encouraged everyone to "get to know your neighbors" (Ooi and Shaw 2004, p. 144).

Beyond the official rhetoric of harmonious togetherness, reserve and a general cautiousness exist beneath the surface. In reality, Singapore society is more of one in which different groups—"CMIO"[3]—share the same space rather than a cultural "melting pot". The existence of groups of "others" within Singapore public space and the involuntary "enclaves" of foreign workers, further complicate any effort to represent Singapore society as a rational and holistic entity. These marginalized groups in the city include those who dwell within the fringe of conservative Singapore society, and those who fall outside of the system rewarded by meritocracy—such as the mentally impaired, physically handicapped, alcoholics, drug addicts, and other social outcasts who remain under-represented in Singapore society or cannot find a way to articulate their needs.

Singapore society is dominated by a highly centralized system of government. The state holds an ambiguous view of other forms of civil society or public activism, as its definitions of activism are limited to voluntary or welfare causes. Forms of public activism existed in Singapore's history, and continue to exist in the contemporary times, but a sustained form of civil society is still in its inchoate stages. The growing civil society is supplemented by discussions in the media (the Forum pages of the Straits Times, for instance), social media, and the rise of the "knowledge economy".

Public participation in policy-making has in the past been limited to focus group discussions, but has developed to include conversations with the state, such as "Our Singapore Conversation" or OSC, and the SGfuture dialogues, held in conjunction with the "Future of Us" exhibition in 2015, to mark Singapore's Golden Jubilee year. The town hall gathering-style OSC was first initiated in August 2012, and had involved over 47,000 Singaporeans participating in over 660 dialogue sessions island-wide over the year-long period. From these Conversations, five core Singaporean aspirations had emerged: Opportunities, Purpose, Assurance, Spirit, and Trust. The SGfuture dialogues, intended to engage Singaporean youths over the nation's future, went a step further in encouraging youths to take action in realizing their visions, with funds provided by the Government. As shared by Minister Chan Chun Seng, one of the lead Ministers of SGfuture: "It's not just about talking and sharing. It's about talking, sharing and, more importantly, doing things together. We must not underestimate *the power of people coming to do things together*" (Straits Times 2016, italics mine).

---

[3]CMIO is often used as a short form for describing the racial groups: Chinese, Malay, Indian and Others.

The form of *public-space-in-practice* described in this study does not depend on talk as the communicative medium, as spatial practice is the modus operandi here. As such, 'discursive public space' within this context has to be defined more broadly than the mode of communicative action found in the Habermasian model. The form of negotiation through spatial practice situated within public space in everyday life, supplemented by discussions in the media such as the Forum page of *The Straits Times*, potentially generates change over time. Negotiations through spatial practice may not be in the form of a 'dialogue', in which channels of communication are open between two parties.

More often than not, spatial practice may influence public opinion or create awareness on the part of the state on the needs of certain groups, trends, resistance, or other forms of knowledge regarding its citizenry. Spatial practice is not only the mode of communication between the people and the state, but also the mode of communication between people. Some form of communication or exchange of information, achieved through spatial and visual means, occurs between different groups who may or may not engage in actual dialogue.

Public-space-in-practice is ultimately the testing ground of state policies, ideologies or space-shaping strategies of the state. As reiterated through the case studies, new ideologies in which to shape the nation are often translated into spatial environments. For example, "Garden City in the Tropics", "Building Communities", "Global City of the 21st Century", and "Communities in Bloom".[4] New spatial paradigms are created in which to socialize its citizens into a way of life or to shape social relations among people. Public spaces are thus the sites of negotiation of these paradigms, when they become *public-spaces-in-practice* and are subject to acts of resistance, transference, transgression or transformation (Fig. 5.1). Spatial practice renders itself as a non-verbal dialectic played out over time and situated within the habituation of everyday life.

As we have established, Singapore public space differs from the classical discursive modes of public space. Yet, we have seen that Singapore public spaces are not less rich or useful as a medium of dialectic between the people and the state, as well as among the constituents of Singapore society. How relevant then is the Singapore model of public space in helping us rethink the existing discourse on public space?

5.2 Rethinking Public Space

The instances of public space reviewed in this study embody concentrated collisions of systems in space—such as that of state control and economic enterprise, of history and cultural systems in changing times, of formal and 'informal' economic actors in space, and of cultural identities being formed in space in relation to others.

[4]This is mentioned in Chap. 2—as part of the "City in the Garden" concept.

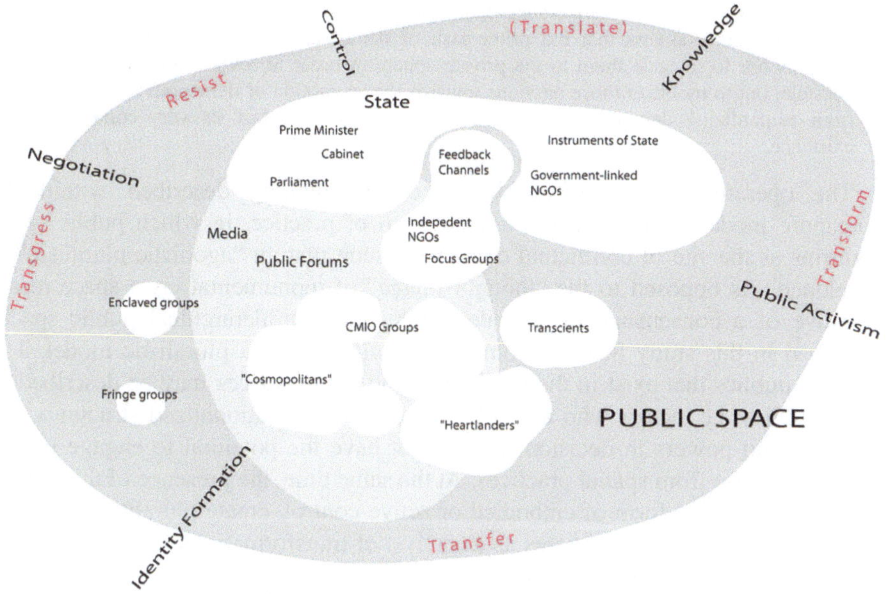

Fig. 5.1 Graphical representation of relations of state, people and public-space-in-practice

The coherent discourse of the bourgeois public sphere described by Habermas does not exist in Singaporean constructions of public spaces, i.e. with mediating civic institutions, the market and the state in dialogue. Although there are many theories of public space, as discussed at the beginning of this study, they do not necessarily reflect one unified conception of public space. However, it is possible to discuss the Singapore model in relation to elements or notions that are usually associated with the discourse on public space, such as democracy, consensus, scale (monumental or otherwise), and transformative potential.

5.2.1 Deliberative Democracy Versus Agonistic Pluralism

The deliberative model of democracy advocated by Habermas necessitates ideal conditions in which communicative action can take place, and requires that differences are bracketed, not highlighted and discussed. As such, his discourse is silent on the construction of identity in public. However, the alternative to collaborative public space is not dissension and chaos—which the Singapore polity posits as the undesirable alternative to its politics of consensus—but works within a framework of pluralism and is in part collaborative and in part conflictual. Such a system, as advocated by theorists like Mouffe (1999), would entail non-hierarchical forms of power, knowledge, and I would include, space:

> Contrary to the model of "deliberative democracy," the model of "agonistic pluralism" that I am advocating asserts that the prime task of democratic politics is not to eliminate passions nor to relegate them to the private sphere in order to render rational consensus possible, but to mobilize those passions towards the promotion of democratic designs. Far from jeopardizing democracy, agonistic confrontation is in fact its very condition of existence.

The operative modes of public-space-in-the-making described within the Singapore model would fall into such a form of practice, in which public space performs as the site of conflictual consensus, such that an "agonistic pluralism" is developed. As opposed to the "thought image" of monumental civic space representative of a consensual, large-scale public, the non-hierarchical public spaces described in this study fit more comfortably within such a pluralistic model. The types of publics that exist in these microcosmic public spaces may be described as weak publics, i.e. actors who are not organized in institutionalized structures and who have no powers in decision making but have the potential to engage in discourse that arise from spatial practices. At the same time, the presence of the state in public space, in the form of embodied or active control, create pluralities in public space as mediator and sometimes as a catalyst of transformations. These pluralities sometimes render space agonistic.

5.2.2 *Public-Space-in-Practice*

The concept of *public-space-in-practice* described in this study is inscribed within a pragmatic model, premised on "things in the making" (Ockman 2000) rather than on the finished product. In other words, the pragmatic *public-space-in-practice* is a working model, operating within and 'trusting' its operative milieu, even if it is critical of the larger context in which it exists.

Modern discourses on public space are based on democracy[5] as a precondition of the existence of a true public sphere. However, as we have established, the political milieu in which public space and the public sphere exist in Singapore does not fall under the ideal western, liberal definitions of democracy. Democracy is arguably 'decentralized' within the Singapore context, such that it exists at local levels of practice. Within each microcosmic system of public space, physical space itself acts as a functional 'democratic resource'. As the 'stage' for enacting social structures and cultural practices, it also exposes tensions and conflicts. In doing so, public space participates in the shaping of democracy, and the notion of public sphere.

As a democratic resource, microcosms of public space enable the formation of identities, representation, the establishment of boundaries and their negotiations. Microcosmic public space, as the physical locus of interaction, allows for the

[5]Democracy is indeed a problematic term as there are many forms of democracies, and the discourse within this realm is plentiful. As I do not intend to engage in a prolonged discourse on democratic forms, Habermas' deliberative form of democracy is referred here.

mixing of races, even under the condition of weak people-to-people relationships. It is a "thing in the making" contributing towards a larger Singaporean identity. Like speech, the media and the Internet, public space is not just the space of communication, but is itself a communicative medium, as it embodies 'knowledge' and 'difference'. As there is communication through public space, there is a potentiality for action, which may lead to consequence and transformation. *Public-space-in-practice* is the transitional medium in which a form of democracy is being molded, even if its final form is yet unclear. Although neither large-scale civic spaces that play the role of democratic public space nor spaces for free speech exist in Singapore, the local-level public spaces as an aggregate serve as testing ground for an adaptive model of public-space-in-the-making.

5.2.3 Communitarian Versus Individual

Instead of evoking spectacular, momentous acts of democracy within transcendental forms of public space, the Singapore model of public space is situated as microcosmic sites within the quotidian rhythms of everyday life. The mechanisms of resistance, transfers, transgression, translation and transformation occur not through the "concerted acts of men coordinated through speech and persuasion",[6] but more commonly through the acts of individuals or small groups of individuals. These are more akin to de Certeau's (1984) and Bourdieu's (1990) concepts of resistance to social practice on the level of individuals. It may seem contradictory that individual action is emphasized, as the Asian trope of community life and shared values remain the dominant rhetoric of the state. However, we have established that Singapore society has low social capital,[7] and the system of meritocracy emphasizes one-upmanship more than communitarianism.

5.2.4 Nostalgia Versus New Sociabilities

Discourses on nostalgia are a function of the rapid development of the built environment of Singapore. The history of public space in Singapore can hardly be cast as a history of decline, since there was a dearth of built environment in its early history, and the public spaces, many which were incidental in early times, were of a quality that was desirable. The objects of nostalgia were really the mimetic functions these

[6]This refers to associative space described by Hannah Arendt in *The Human Condition* (1958), as opposed to the agonistic space of appearance that she also describes.

[7]This is generally the case, however the Malay community, as a minority group, has strong ties within.

public spaces performed, as their loss signified the demise of natural spaces, or simpler lifestyles, or the existence of the more boisterous street life of old.

The decline in the importance of what would qualify as "civic spaces" in the city, such as the Padang, is also informative of the trajectory in which public space has developed. While the central location and visibility of the Padang made this the site of many protests and rallies in the past, its impact as the representative space of the city has been reduced by the decentralization of Singapore and the ascendance of the media taking over the role of public space. When the worldwide Millennium Countdown was broadcast, it was Orchard Road that was selected as the significant public space of the city, not the Padang. The legitimacy of Orchard Road as the public space of the nation lies in its ability to flow with the times and constantly re-invent itself. New sociabilities are created in space when personal memories in public space are documented and re-enacted over time, eventually becoming embodied within notions of public space.

Concepts not traditionally associated with classical traditions of public space, such as recreation and consumption, the latter of which is more often associated with the decline of public space, could be considered as ways of engaging with public space within the Singapore context, as public spaces also function as spaces of sociabilities. Practices occurring within the realms of recreation and consumption constitute part of spatial practices that have space-shaping potentials. As Sharon Zukin had noted, people are more active as consumers than as citizens (1996), a case that is very much applicable to Singapore public space.

5.3 Trajectories: Aspects of Singaporean Life with Possible Impacts on Public Space

Aspects of Singaporean life in public space have been changing over the years. The Singapore we are currently investigating is one that is negotiating a multitude of changes to the makeup of its population. As outlined in the earlier chapters, Singapore is home to an aging population which has a low fertility rate. This reduces the total factor productivity in Singapore which renders the domestic workforce weak and reliant on foreign labour. Today, foreign workers have come to represent almost a third of Singapore's workforce. This influx of foreigners has created friction among Singaporeans, especially in terms of the sharing of public space and amenities, such as transportation systems. Furthermore, as Singapore opens up its labour markets to attract global workers at differing scales of the workforce spectrum, issues of income difference become more stark. These trajectories of urban life need to be accounted for as we encounter new paradigms of public space. The Singapore of the future needs to design for differences and diversity. As such, how can designers contribute to new imaginations of public space, based on the information gathered from this research, and from the new anticipated trajectories of the "re-making" of Singapore to face a more globalized and competitive world? Some possible design futures of public space would be discussed under broad themes that have become apparent from this research.

5.4 New Paradigms in Singapore Public Space

5.4.1 Space-Time Dimension

The notion of public space being of a specific location and particular typology is no longer the only form of public space. Instead public space, such as the previously discussed "-scapes", are constantly in flux, and are sometimes experiential, i.e. experienced in a sequence, instead of a defined form. Public space tends to be more temporally situated, existing only within certain time cycles or time frames, or consumed within "public time". They are experienced within a time period shared by a good proportion of "public". Public space need not only be designed as concrete space, but it can also be anticipated as experienced space within time-space patterns, which may be overlapping and complex. With the push towards live-work-play environments in Singapore, public space would allow for even more overlapping fields of influence and complex rhythmic patterns of use instead of being mono-zoned, as in the past.

While there is a move away from iconographic spaces with commonly held meanings, there is a danger of designers perpetuating generic, bland spaces that are indistinguishable from others. There is a need to retain the potentials of contested meanings within public space. The aesthetics of space design, although difficult to prescribe for dynamic place-scapes, risk being complicit in making space undifferentiated in its embrace of greater inclusivity in design. The performative aspects of public space design, such as the use of materials, appropriate scale and detailing may not seem highly significant. Yet they leave their mark in the perceptions of those who engage with the spaces—and in the poetic imagination, as we can see from this excerpt of "Stone Table" (2004) from Alfian Sa'at:

> Even as the void deck is a space of relentless transit, there is a kind of touchstone represented by the stone table, a space which invites rest. Around the table are six cylindrical stone chairs, and on the table surface itself is carved the board layout to a game of either Chinese chess or draughts. The fact that such public furniture is made of stone invests it with a sense of the mythic, with the permanence of statue and gravestone…
>
> The austerity of the stone table has not repelled the young; conversely the juvenile graffiti on its surface has not deterred the old from utilising it as temporary refuge. In upgraded estates, however, the stone chairs have been replaced by plastic ones, whose perforated backings are designed to thwart attempts at graphic vandalism. These new furniture are probably as resistant, if not more, to wear and tear. And yet there remains a distinction to be made between the ageless and the non-biodegradable.

5.4.2 Spaces of Exchange

With an increasingly globalized workforce within the Singapore economy, public space in the city would host even more diverse entities, conflating different identities and spatial associations within more concentrated spaces. The new compression of spaces would have to be imagined with new forms of connections and "connectivities" between these dense networks of spaces. Devices of boundaries, seams and layering over of spatial fields have to be employed to deal with the needs of different groups of people in space.

In order to 'enable' new sociabilities in public space, these public spaces should not be conceived as homogenous and seamless spaces. These should instead embody zones of particularity, and even territories open to contestations in order for "friction" to occur. In other words, space should possess both the elements of "smooth" and "striated" space, with opportunities for space to be claimed and appropriated, allowing for improvisation instead of over-programming. The new geographies of public space would be described more as mosaics and networks than as seamless interfaces.

5.4.3 Space Menus

Public space, although a shared resource and common ground, need not be common, but it should be malleable within personal menus of *space-a-la-carte*. Opportunities for meeting others, and those in which one can inhabit with other "insiders", would provide a range of diverse spatial experiences as well as familiar spaces of interaction. Spaces that have functions as transitory spaces and interstitial spaces, such as "non-places", might be exploited for these functions as collective spaces, by integrating these with other networks of public spaces. Such public spaces espouse temporary communities within tight proximities, engendering opportunities to recognize "others" and to partake in "public time".

Atrium spaces within buildings, and spaces in-between "outside" and "inside" could be conceived as "big city rooms" and "plug-in" spaces due to their accessibility from streets and other public spaces. Designing for flux enables public spaces to intrinsically embody elements which are fixed and those which are changeable, so that spaces are experienced differently at different times, or have the ability to morph over time. People in public spaces may be incorporated within space's ability to be changed, so that human agents become complicit in the vicissitudes of public space. However, with the proliferation of privately owned and developed public spaces, it might become necessary to devise development policies that still make these places available for public use without becoming "paid-for" spaces which are programmatically fine-tuned for the sole purpose of consumption.

5.4.4 *New Technologies in Space*

Materially, the architectural substance defining spaces is shifting from solid building forms and shapes, to new typologies predicated on surfaces, such as video-walls, changing seasonal facades, and even morphing walkway paving patterns in a constant updating of images. The new ephemerality and instability of appearances are concomitant with the displacement of bodies in city spaces, embodying shifts in affinities to spaces. Even the forms of representation of places in the city have changed from traditional figure-ground maps and photographs, to abstract forms like graphs anticipating real-estate and rental values, demographic patterns and scopes of surveillance such as camera-surveyed atriums, MRT stations and sidewalk bank automats and Electronic Road-Pricing (ERP) gantries.

The overlay of social media in the urban landscape—from apps that help with transportation options, choices of where to eat, visit, and even who to befriend in the city—makes the invisible interactions of people and places even more complex. These complex array of technologies in spaces are complicit in new ways by which public spaces and urban phenomena are perceived—the use of mobile phones can determine at will spaces for rendezvous in the city, or remove individuals from the physical spaces that they inhabit temporarily. Animation of public spaces with the use of digital displays or interactive features could serve to add textures to the perception of these spaces, or even gather people in public as participants within digitally-induced lights and sounds in public space. The increased use of Internet-powered information kiosks in public spaces facilitates public space's function of dispensing information in ways never imagined before.

5.4.5 *Home-Ground*

Spaces near the home, rather than being overly "contained" and "territorized", should be arranged within a less hierarchical form of spatial networks constituted through channels and nodes. These have been found to work best when habituated within the routines of everyday life. The nature of spaces within public housing in the long term should perhaps be conceived to be integrated with retail and ancillary institutions in the new town. These facilities can then be part of the networks of public spaces, rather than concentrated within calculated zones of efficiency within the new towns.

5.4.6 *Degree of Control*

Although some degree of both active and passive forms of control are necessary for the safe functioning of public spaces, such as rules of acceptable behavior and the need to manage conflicts in public space, too many inhibitions in public space, such as the overusage of surveillance technologies or barriers to accessibility, would not be conducive to creating convivial public space. On the other hand, devices of control, such as gantries at the entrances to public libraries, serve as a reminder to denote a crossing into zones where certain public decorum is expected, without overtly interfering with accessibility to the libraries. Control in public space may even play facilitating roles, such as providing a peace of mind or to prevent unruly behavior, or even proactive control that anticipates needs before they arise—acting almost like a feedback loop between the state and the people.

5.4.7 Design Framework

The trajectories of Singaporean politics, cultural and social life as well as the urban dynamics of the emerging global city-state seem to suggest that Singaporean public space would move towards more concentrated experiences within the urban realm, and offer a greater variety of choices and encounters in the foreseeable future. Design frameworks which engage heterogeneity in space, encourage adjacencies, manage complexities and are open to change without giving into "anything goes" resignations (Pollak 2004, p. 83) will have to come through new collaborations from design consultants with public and private interests, and with the recognition of the innate dynamics already operating in public space. Public space in the city can no longer be considered as singular spaces but as a complex weave of social fields.

It may come as no surprise that some of the trajectories of public space development in Singapore may appear to be applicable to other globalized cities in the world, and are not particular to Singapore. The spatial practice dynamics between the state and the people, and among different groups in free space may stem from local histories and specificities. However, the development of new paradigms in physical space takes on possible trajectories that would be applicable wherever good public space is desired.

5.5 Futures and Implications for Design

How the state and its agents, as well as Singaporeans, imagine public space to be would change with the dynamic trajectories of Singapore's growth. In what ways and forms can architects, planners and policy makers respond to these trajectories? What kinds of new imaginations of public space would emerge from responding to the new discourses in political, social, and cultural milieus? What is the role of design and designers operating within these emergent discourses and imaginations?

Up to the 1990s, the approach to public space provision in Singapore had been one predicated on 'aestheticizing' the environment—such as the use of landscapes and new finishes to improve the appearance of places, reduce crime, or most recently, the "theming" of places—such as the renovated Chinatown overpass themed with Chinese pavilions and motifs. The critique of displacing content with aesthetics was most vociferous with the Singapore Tourist Board's Plan to thematically re-invent Chinatown, complete with a built-in *wayang* stage in the main square of the ethnic quarter.

New collectives of *chi-chi* public spaces, catering only to a narrow slice of the demography—typically the yuppie crowd and the expatriate market—have sprung up in the areas around the historic districts like Tiong Bahru, the revitalized Singapore River, such as in and around Robertson Quay, colonizing yet again the historic River. The argument is not so much about "authenticity" versus "nostalgia", but more so about the programming of spaces which are colonized by analogous groups, leading to "frictionless" and exclusionary public space.

On the brighter side, there is a definite "spatial turn" to Singapore planning from the 1990s onwards, with the conservation of whole historic districts instead of isolated buildings or city blocks. Public spaces are taken into account in these Masterplans, as well as the consideration of districts as a whole in the development of new planning guidelines. Streets such as Orchard Road are considered as a holistic space in the "Make Orchard Road More Happening" strategies in the early 2000s and the Remaking Orchard Road initiatives that followed. In the Concept Plan 2001 proposals as well as the *Public Spaces and Urban Waterfront Master Plan* launched in 2003, public space is given more importance, especially with integrated zoning plans for live-work-play environments.

Judging from trajectories in political, cultural and social aspects of Singaporean life, public space would occupy an important position in new Singapore. There would be an increased investment of political capital within public space to ensure more social mixing of the races and better social cohesion between "Cosmopolitans" and "Heartlanders", as well as to bring to the open, discussions of political and social issues in the interest of maintaining a stable, multiracial society. Public spaces would also become the sites of artistic and cultural expression, as they are an effective medium of transmitting the arts to Singapore society and to the global cultural landscape. Public spaces would also be one of the arenas in which the inchoate civil society and public activism could be nurtured. They could form an important channel of communication between the state and the people, as well as within the pluralistic entities of Singaporean society itself.

Although the state, through its instruments such as the Urban Redevelopment Authority and the Singapore Tourism Board, had been the dominant provider of public space, increased collaboration with private agencies such as developers, building owners, arts groups, and civil society groups, as well as with actors such as planners, architects, artists and educators, would be foreseeable in the future. The more collaborative climate of public space production will come in anticipation with the divesting of state interests among autonomous groups and individuals, as such initiatives come with the maturing of the economy and the arts and culture development. Such a shift is already evident, for example, in the development of the Singapore River Promenade, in which the URA worked in concert with private developers, who leased land along the riverbanks. The conceptualization of public space along the promenade involved businesses, arts groups as well as different semi-autonomous agencies of the state. The place-making efforts at the Singapore River are undertaken by Singapore River One (SRO), a private sector-led partnership charged with the day-to-day management, maintenance, enhancement and marketing of the Singapore River.

Singapore's striving to take on a new identity as a 'Smart Nation' has led to its transition from a knowledge-based economy to a bastion of innovation as a 'Global-Asia Hub'. Since the 1980s, Singapore has invested in biotechnology and has created 'one-stop' hub of industrial centers for biomedical firms looking to establish themselves in the region. Singapore has always prided itself as an innovation economy, as it continuously invests funds in scholarships and fellowships to attract new talent, spur innovation and promote interdisciplinary, public-private collaboration between sectors. However, Singapore's quest to become a Smart Nation is not just one which boasts smart jobs, but one that takes advantage of technological advancements in order to enhance its urbanisation and place-making processes. From the individual's use of popular smart phone apps to share experiences and stories of public space in Singapore, to the highly sophisticated water sensor technologies that collect data on water levels in drains and canals at ten minute intervals to provide flash flood information to users via a SMS alert system they can subscribe to,[8] Singaporeans' access to (and understanding of) data is crucial in determining if it can earn the label of a 'Smart Nation'.

In addition, Singapore is oft heralded as one of the best case studies of a biophilic city—a city that fosters biodiversity and connections to nature within its urban environs (Beatley 2011). The notions of greening and biodiversity are interwoven within Singapore's urban planning experience in creating a city of gardens and water. Even as Singapore is well established as a garden city in the tropics, Singapore's existing 32 rivers, 17 reservoirs and 8000 km of waterways have all been key assets in its "blue" makeover (Khoo 2016). When touring the ABC Waters (Active, Beautiful and Clean Waters)[9] exhibition in 2007, Mr. Lee Kuan Yew declared, "Momentum is growing, as people start to see the value of these projects. You will have the support of the people in this programme as they will very soon realise that their property value will increase if there is an ABC Waters project next to their home".

While Singapore's greenery serves a very practical function of beautifying the city and acts as a mitigator of the urban heat island effect, a biophilic city is more than one that incorporates nature in its surroundings—it is one that fosters an acute

[8]See PUB's fact sheet the use of water level sensors/CCTVs to manage flash floods. http://www.pub.gov.sg/managingflashfloods/WLS/Pages/WaterLevelSensors.aspx.

[9]The Active, Beautiful, Clean (ABC) Waterways project is an initiative to capitalize on Singapore's extensive water catchment networks, which amount to two-thirds of Singapore's land area. The project integrates drains, reservoirs, and canals to create a system of waterways which complement public park space, resulting in picturesque community spaces for exercise and recreation.

awareness of the multiple ways in which one interacts with the environment, and promotes a shared environmental consciousness among users of public space. So, as Singapore pursues its other environmental objectives, such as promoting a 'cultural shift' towards a 'car-lite' future (Straits Times 2016), it needs to ensure that environmental education, heritage preservation, green design, and community investment all come together in a compelling future of biophilic Singapore which stakeholders can embrace as a win-win situation for all.

5.6 Conclusion: The Singapore Model in Transition

This study has shown the limitations of applying the western conception of public space through its discussion of the operative model of Singapore public space. Models which only recognize the democratic agenda of public space may actually be missing alternative dynamics through which public space thrive. The Singapore Model of public space is an interpretive model, grounded within its specific context and history. The model is by no means an ideal, but it provides an alternative to classical models based on the pre-condition of ideal western liberal democracy. The Singapore model offers spatial dialectics rather than talk as the medium of communication, and functions within a milieu, which is vastly different from those within western democratic systems. The findings suggest that the discourse on public space as understood in the ideal western democratic tradition, when applied to Singaporean public space, offers much criticism but is inadequate in identifying alternative processes that allow for transformative potentials in public space.

The Singapore Model is an adaptive model within such a milieu, and could be a frame of reference for public spaces within similar systems, where a strong, dominant state, weak civil societies and pluralistic entities exist, so that rational-critical debate between equals cannot be a precondition of its formation. The pluralistic nature of Singapore public space is contrary to the seemingly homogenous image of an Asian society purported to have Shared Values. While strong communities exist in Singapore, particularly within racial groups, weak links between these result in a lack of social cohesion. Many of the spatial practices in public space involve individual actors or small groups, rather than large scales of public. Nostalgia, a phenomenon stemming from the rapid pace of Singapore's urban development, is present but is not an overriding factor in the production of Singapore public space, which tends to find legitimacy in new sociabilities than in times past.

The model is also an emergent one, constantly in a state of making, as the trajectories within its milieu change. Singapore public space is intrinsically spatialized, as political, social and cultural life are explicably spatial, and do not remain within the abstract imagination. Just as these milieus are spatialized, so, too, is the spatial realm impacted by political, social and cultural factors. The trajectories through which Singapore develops would no doubt impact on the future of its public space.

As mentioned in the first chapter, the 2011 General elections brought about a 'new normal' for Singaporean politics, resulting in a more active citizenry which opened up a democratization of decision-making processes concerning Singapore's development and future. This 'new normal' has coincided with suggestions by local political leaders on whether Singapore is in danger of entering a 'mid-life crisis' (Vietor and White 2013). The late Mr. Lee Kuan Yew was also wary of Singapore's future, and wrote in August 2013: "Will Singapore be around in 100 years? I am not so sure … I am absolutely sure that if Singapore gets a dumb government, we are done for" (The Sunday Times 2013). In order for the government to avoid this ignominy, it must be one that adapts with the changing needs of its people. A Singapore model in transition requires a government that is anticipatory and pro-active; one that engages community participation in the process of spatial planning and the manufacture of public space. This is the remedy that will prevent Singapore from falling into the malaise of a 'mid-life crisis' and uncertainty around the 'new normal'. Rather, Singapore's half-century of independence and political 'new normal' should be embraced as a coming of age, and its newfound citizen engagement should be harnessed for broad-based policy-making within public spaces. Singapore now has an opportunity to receive and act on feedback to "remove policies that artificially entrench or perpetuate existing spatial inequalities" (Urban Solutions 2014), improve access to common infrastructure and amenities, and promote co-mingling in shared spaces in Singapore.

Even scarcer than space in Singapore is public space, with complex capillaries of rules, customs, and norms governing its use, access and practices. Public space will continue to be a point of contention as Singapore's population expands. Thus far, the production of public space in Singapore has been a man-made construct—a concerted planning outcome of planning authorities and political leaders that is historically devoid of public engagement. However, as Singapore comes of age, its social makeup is undergoing a significant change. Its evolving society has now become more politically engaged, especially through the deployment of social media platforms. This has resulted in a re-valorization of social values and practices. Singapore must have the political wherewithal to embrace this change, and its strategies for moving forward must allow for the construction of public space to be a democratizing force that allows for these spaces to be the product of both state-level development goals and the needs and wants of society. A biophilic, smart Singapore will remain just another 'adjectified city' without buy-in from residents, or if they are unequipped with the know-how on how to take advantage of these upgrades.

The future of public space will be one that is more scrutinized, surveyed, and contested than ever before. For Singapore to continue to thrive, it must take advantage of this opportunity to bring together multiple actors and create stakeholders in public space so that the public-ness of public space is preserved.

Glossary

Central Business District (CBD) The core financial and business district of Singapore, located south of the Singapore River. The area encompasses Orchard, Chinatown, Bugis, City Hall, Raffles Place/Tanjong Pagar and Nicoll Highway. Further development of the CBD is projected to take place in Marina Bay

Central Provident Fund (CPF) The mandatory social security savings scheme for Singaporeans and Permanent Residents, instituted by the Ministry of Manpower

City in a Garden Singapore's campaign and vision to be a green city introduced in 2013, building upon its Garden City status. Its goals include engaging communities to create a green Singapore, building the Singaporean landscape and horticulture industry, enriching biodiversity in the environment, establishing world class gardens, optimizing urban spaces for greenery and recreation and rejuvenating urban parks and enlivening streetscapes

Developmental city-state The condition where the state (in Singapore's case—city-state) establishes its legitimacy to govern upon its ability to generate and maintain economic growth

Garden City Vision officially introduced by then Prime Minister Lee Kuan Yew in 1967 to transform Singapore into a city with abundant and lush greenery that would create a pleasant urban living environment for Singaporeans, as well as attract foreign investment to the city-state

Housing Development Board (HDB) The statutory board under the Ministry of National Development that is responsible for resettlement and the provision of public housing

Kampong The traditional rural Malay village that was the main form of housing for most Singaporeans before the mass provision of public housing by the Singapore Improvement Trust (SIT) and the Housing Development Board (HDB)

Lee Kuan Yew Widely regarded as the founding father of postcolonial Singapore, Lee Kuan Yew was the founder of the ruling political party of Singapore, as well as the first Prime Minister of Singapore from 1959 to 1990. He subsequently became Senior Minister from 1990 to 2004, and Minister Mentor from 2004 to 2011.

L. Hee, *Constructing Singapore Public Space*, Advances in 21st Century Human Settlements, DOI 10.1007/978-981-10-2387-3

Mass Rapid Transit (MRT) The MRT is a rapid transit system that spans 170.7 km across Singapore with 101 stations. Opened on 7 November 1987, the MRT has expanded to become an integral mode of public transportation for Singaporeans.

Meritocracy The ideology that Singaporeans should be rewarded on the basis of competence and not on aspects of racial background or identity

Multiracial Society Concept of a society that is characterized by the peaceful coexistence of various races, languages and religions

National Parks Board (NParks) The statutory board under the Ministry of National Development responsible for managing and enhancing the greenery of Singapore, as well as conserving Singapore's biodiversity

Non-place A concept by anthropologist Marc Augé to describe a interstitial space where one passes through. The meaning of the space is derived from the activities one conducts in it, instead of its status as place.

Peoples' Action Party (PAP) The ruling political party in Singapore that has won every election since the first general election in 1959. It was founded by Lee Kuan Yew in 1954

PubliCity A URA initiative launched in the draft Masterplan of 2013, centered on public engagement to celebrate public spaces, and to enliven public space through design and program interventions in Singapore

PUB (Singapore's National Water Agency) Statutory board under the Ministry of Environment and Water Resources. It is the national water agency that manages Singapore's water supply, water catchment and water use.

Shared Values Confucianist values that have been adapted into Singapore's national ideology in 1991. The four core values were, (1) community over self, (2) upholding the family as the basic building block of society, (3) resolving major issues through consensus instead of contention: (4) stressing racial and religious tolerance and harmony.

Singapore Improvement Trust (SIT) The Housing Development Board's predecessor set up in 1927 by the British colonial government to provide affordable public housing

Smart Nation Initiative to attract capital, talents and ideas on Information and communications technology (ICT) to Singapore, aimed at allowing people to "live meaningful and fulfilled lives, enabled seamlessly by technology" (Lee 2014)

Urban Redevelopment Authority (URA) Statutory board under the Ministry of National Development that is responsible for land planning, development control, and urban conservation in Singapore. Founded in 1974, it was initially instituted as the Urban Renewal Department (URD) in 1966.

Void deck The ground level of the HDB flat, designed to be an open space for multiple uses. Events that are held at the void deck have to be sanctioned by the HDB town council: and they regularly include funeral wakes and weddings.

Bibliography

Abraham A (2003) Awestruck on Orchard road. EastWest Books, Aminjikarai
Abrams C, Kobe S, Koenigsberger O (1963) Growth and urban renewal in Singapore. UN Technical Assistance Department of Economic and Social Affairs, New York
Appadurai A (1996) Modernity at large: cultural dimensions of globalization. University of Minnesota Press, Minneapolis
Arendt H (1971) The human condition. University of Chicago Press, Chicago
Arendt H (1992) Walter Benjamin: 1892–1940. In: Arendt H (ed) Illuminations. Fontana Press, London, pp 7–58
Augé M (1995) Non-places: introduction to an anthropology of supermodernity. Verso, New York
Barr MD (2000) Lee Kuan Yew: the beliefs behind the man. Curzon Press, Richmond
Baudrillard J (1994) Simulacra and simulation (trans: Glaser SF). University of Michigan, Ann Arbor (French edition: Baudrillard J (1981) Simulacres et Simulation. Editions Galilée, Paris)
Beatley T (2011) Biophilic cities: integrating nature into urban design and planning. D.C., Island Press, Washington
Beck U (1986, 1992) Risk society, towards a new modernity (trans: Ritter M). London, Sage
Benhabib S (1992) Models of public space: hannah arendt, the liberal tradition and jürgen habermas. In: Calhoun C (ed) Habermas and the public sphere. MIT Press, Cambridge, pp 73–98
Benjamin W (2002) Das passagen-werk. Suhrkamp Verlag, Frankfurt. (English edition: Benjamin W (2002) The arcades project (trans: Eiland H, McLaughlin K). New York, Belknap Press)
Bishop R (2004) Beyond description: Singapore space historicity. In: Bishop R, Phillips JW, Yeo WW (eds) Beyond description: Singapore space historicity. Routledge, London, pp 1–17
Bourdieu P (1990) The logic of practice. Stanford University Press, California
Bourdieu P (1977) Outline of a theory of practice (trans: Nice R). Cambridge University Press, Cambridge
Borja J, Castells M (1997) Local and global: management of cities in the information age. Earthscan Publications, with the United Nations Centre for Human Settlements (Habitat), London
Branegan J (1993) Is Singapore a model for the west?. Time Magazine
Buck-Morss S (1991) The dialectics of seeing: Walter Benjamin and the arcades project. MIT Press, Cambridge
Caldeira T (2000) City of walls: crime, segregation, and citizenship in Sao Paulo. University of California Press, Berkeley
Castells M (1992) Four asian tigers with a dragon head: a comparative analysis of the state, economy, and society in the Asian Pacific Rim. In: Appelbaum RP, Henderson JW (eds) States of development in the Asian Pacific Rim. Sage, Newbury Park
Chai KW (2001) Political climate control on high. Far East Econom Rev, 22 March 2001
Chang TC (1996) Local uniqueness in the global village: heritage tourism in Singapore. Dissertation, McGill University.

L. Hee, *Constructing Singapore Public Space*, Advances in 21st Century Human Settlements, DOI 10.1007/978-981-10-2387-3

Choe AFC (1969) Urban Renewal. In: Ooi JB et al (eds) Modern Singapore. University of Singapore, Singapore, pp 165–166
Chua BH (1997) Political legitimacy and housing: stakeholding in Singapore. Routledge, New York
de Certeau M (1984) The practice of everyday life (trans: Rendall S). University of California Press, Berkeley
de Certeau M, Mayol P (1998) the practice of everyday life: living and cooking. University of Minnesota Press, Minneapolis
Deutsch R (1992) Art and public space: questions of democracy. Soc Text 33:34–53
Deutsche R (1996) Evictions: art and spatial politics. MIT Press, Cambridge
Deutsche R (2000) Democratic public space. In: Ockman J (ed) The pragmatist imagination: thinking about "Things in the Making". Princeton Architectural Press, New York
Federal Research Division, Library of Congress (1989) Country studies: Singapore, Dec 1989. http://lcweb2.loc.gov/frd/cs/sgtoc.html
Fiske J (1992) Cultural studies and the culture of everyday life. In: Grossberg L, Nelson C, Triechler P (eds) Cultural studies. Routledge, New York, pp 154–174
Foucault M (1977) Discipline and punish: the Birth of the prison (trans: Sheridan, A). Pantheon Books, New York.
Foucault M (1977) Power/knowledge: selected interviews and other writings. Pantheon, New York
Foucault M (1984) Of other spaces. Architecture/Mouvement/Continuité
Fraser JM (1948) the work of the singapore improvement trust, 1927–1947. Authority of Singapore Improvement Trust, Singapore
Fraser N (1990) Rethinking the public sphere: a contribution to actually existing democracy. Soc Text 25(26):56–80
Garner RE (1972) The politics of urban development in Singapore. Cornell University Press, London
Gastil RW, Ryan Z (2004) Introduction. In: Open: new designs for public space. Van Alen Institute, New York
Gibson W (1994) Disneyland with the death penalty. Wired Mag 1:4
Giddens A (1990) The consequences of modernity. Stanford University Press, Stanford
Gopalkrishnan V, Perera A (1983) Singapore: changing landscapes—Geylang, Chinatown, Serangoon. FEP International, Singapore
Habermas J (1989) The structural transformation of the public sphere: an inquiry into a category of bourgeois society (trans: Burger T). MIT Press, Cambridge
Hajer M, Reijndorp A (2002) In search of new public domain: analysis and strategy. NAi Publishers, Rotterdam
Han FK, Fernandez W, Tan S (1998) Lee Kuan Yew: the man and his ideas. Singapore Press Holdings/Times Editions, Singapore
Hee L, Ooi GL (2003) The politics of public space planning in Singapore. Plann Perspect 18 (1):79–103
Highmore B (2002) Everyday life and cultural theory: an introduction. Routledge, London
Joseph I (2000) Reconsidering pragmatism and the Chicago school. In: Ockman J (ed) The pragmatist imagination: thinking about "Things in the Making". Princeton Architectural Press, New York, pp 186–195
Khoo TC (2016) City of gardens and water. Urban Solutions 8:54–63
King AD (1985) Colonial cities: global pivots of change. In: Ross R, Telkamp GJ (eds) colonial cities: essays on urbanism in a colonial context. Martinus Nijhoff, Dordrecht
Koh BS (1992) A brief history of Toa Payoh. In: A brief history of Toa Payoh and other poems. Imperial Publishing House, Singapore
Koolhaas R (1996) Conversations with students. In: Kwinter S (ed) Princeton Architectural Press, New York
Kong L, Yeoh BSA (2003) The politics of landscapes in Singapore: constructions of 'Nation'. Syracuse University Press, Syracuse

Kraar L (2004) The son also rises can Singapore's new leader step out of his father's shadow—and keep one of the world's success stories going? Fortune Mag 150(4):34–42

Lai AE (1995) Meanings of multiethnicity—a case study of ethnicity and ethnic relations in Singapore. Oxford University Press, Kuala Lumpur

Lefebvre H (1991) The production of space. Blackwell Publishers, Oxford

Liu TK (1975) Design for better living conditions. In: Yeh SHK (ed) Public housing in Singapore—a multi-disciplinary study. Singapore University Press for HDB, Singapore

Liu TK (1982) A review of public housing in Singapore. In: Jayakumar S (ed) Our heritage and beyond—a collection of essays on Singapore, its past, present and future. National Trades Union Congress, Singapore, p 134

Liu TK (1992) Public housing in Singapore: an architectural portrait. In: Proceedings of housing design 2000 conference, Singapore, Sept 1992

Lofland LH (1973) A world of strangers: order and action in urban public space. Basic Books, New York

Lofland LH (1998) The public realm: exploring the city's quintessential social territory. Walter de Gruyter, New York

Low S (2000) On the plaza: the politics of public space and culture. University of Texas Press, Austin

Mouffe C (1999) Deliberative democracy or agonistic pluralism? an agonistic model of democracy. Prospects for Democracy 66(3): 745–758

Mouffe C (2000) For an agonistic public sphere. In: Ockman J (ed) The pragmatist imagination: thinking about "Things in the Making". Princeton Architectural Press, New York, pp 66–75

Naidu D (1999) Sungei road: a market on the margins. Singapore Architect 204:134–143

National Parks Board. Community in bloom initiatives. https://www.nparks.gov.sg/gardening/community-in-bloom-initiative. Accessed 23 June 2016

National Parks Board. Speakers' corner. https://www.nparks.gov.sg/gardens-parks-and-nature/parks-and-nature-reserves/hong-lim-park. Accessed 23 June 2016

National Population and Talent Division (2013) Population white paper 2013: a sustainable population for a dynamic Singapore. http://population.sg/whitepaper/downloads/population-white-paper.pdf. Accessed 23 June 2016

Ockman J (2000) The pragmatist imagination: thinking about "Things in the Making". Princeton Architectural Press, New York

Ong AW (2004) Intelligent island, baroque ecology. In: Bishop R, Phillips WJ, Yeo WW (eds) Beyond description: Singapore space historicity, p 176–190

Ooi GL, Shaw B (2004) Beyond the port city: development and identity in 21st century Singapore. Singapore, Prentice Hall

Ooi GL, Tan TW (1992) The social significance of public spaces in public housing estates. In: Chua BH et al (eds) Public space: design, use and management. Singapore University Press, Singapore

Orlikowski WJ (2000) using technology and constituting structures: a practice lens for studying technology in organizations. Organ Sci 11(4):407–408

Paetzold H (2004) Post-functionalist urbanism, the postmodern and Singapore. In: Bishop R, Phillips JW, Yeo WW (eds) Beyond description: Singapore space historicity. Routledge, London, pp 146–165

Perry M, Kong L, Yeoh BSA (1997) Singapore: a development city state. John Wiley and Sons, UK

Pollak L (2004) Paradoxical spaces. In: Gastil RW, Ryan Z (eds) Open: new designs for public space. Van Alen Institute, New York

PUB (2014) 'Active, Beautiful, Clean (ABC) Waters' design guidelines. https://www.pub.gov.sg/abcwaters/Documents/ABC_DG_2014.pdf. Accessed 23 June 2016

Pugh C (1983/1984) The political economy of housing in Singapore. Staff seminar paper no. 13, Department of Economics and Statistics, National University of Singapore, Singapore

PuruShotam N (2003) Silent witnesses: the 'Woman' in the photograph. In: Chan KB (ed) Past times: a social history of Singapore. Singapore, Times Edition

Quah JS (ed) (1990) In search of Singapore's national values. Institute of Policy Studies and Times Academic Press, Singapore

Rajchman J (1997) A new pragmatism? In: Anyhow. Netherlands Architecture Institute, Rotterdam

Sa'at AB (1998) Void deck. In: One fierce hour. Landmark Books, Singapore

Sa'at AB (2004) Stone table. Unpublished excerpt.

Sassen S (1998) Introduction: whose city is it? Globalization and the formation of new claims. In: Globalization and its discontents. The New Press, New York

Sennett R (1977) The fall of public man: on the social psychology of capitalism. Random House, New York

Siddique S, PuruShotam N (1990) Singapore's little India—past, present and future, 2nd edn. Singapore, Institute of Southeast Asian Studies

Singapore Parliament. Liquor control (Supply and Consumption) bill. https://www.parliament.gov.sg/sites/default/files/Liquor%20Control%20(Supply%20and%20Consumption)%20Bill%201-2015.pdf. Accessed 23 June 2016

Singapore Tourism Board (2002) Little India: ethnic quarter. Singapore, Singapore Tourist Board

Sinha V (2003) Merging 'different' sacred spaces: enabling religious encounters through pragmatic utilisation of space? Contrib Indian Sociol 37(3):459–494

Soja EW (2000) Postmetropolis: critical studies of cities and regions. Blackwell, Malden

Sorkin M (ed) (1992) Variations on a theme park. Hill and Wang, New York

Stallybass P, White A (1986) The politics and poetics of transgression. Cornell University Press, Ithaca

Tamney JB (1996) The struggle over Singapore's soul: western modernization and asian culture. Walter de Gruyter, Berlin

Tan MCC (2016) Performative silence: race, riot and the end of multiculturalism. J Interdisc Voice Stud 1:9–22

Tan TKJ et al (1985) Physical planning and design. In: Wong A, Yeh SHK (eds) Housing a nation: 25 years of public housing in Singapore. Maruzen Asia for HDB, Singapore

Teo SE (1996) Character and identity in Singapore new towns: planner and resident perspectives. Habitat Int 20(2):279–294

Urban Redevelopment Authority (1995) Historic district: Little India. Singapore, Urban Redevelopment Authority

Urban Redevelopment Authority (1998) Punggol planning area—planning report 1998. Singapore, URA

Urban Redevelopment Authority (2000) Concept plan review: focus group consultation—final report on land allocation 2000. Singapore, URA

Urban Redevelopment Authority (2015) Streets for people how-to guide. https://www.ura.gov.sg/MS/CarFreeZones/Events/Streets-for-People/~/media/CarFreeZone/Streets_For_People/PDF/Streets_For_People_How-to_Guide.ashx. Accessed 23 June 2016

Urban Redevelopment Authority (2014) URA opens registration for PARK(ing) day. https://www.ura.gov.sg/uol/media-room/news/2014/aug/pr14-47.aspx. Accessed 23 Jun 2016

Vietor, RHK, White H (2013) Singapore's "Midlife Crisis"? Harvard Business School Case 714–039.

Woon W (1991) Commentary. In: Tan TL (ed) Singapore: the year in review 1990. Times Academic Press, Singapore, pp 16–28

Yeoh B (1996) Contesting space: power relations and the urban built environment in colonial Singapore. Oxford University Press, New York

Yeh SHK et al (1972) Homes for the people: a study of tenants' views on public housing in Singapore. Government Printing Office, Singapore

Yeo TCS (1999) Socioscapes in a global city: Singapore's Little India. Dissertation, National University of Singapore

Yeo WW (2004) Of trees and the heartland. In: Bishop R, Phillips JW, Yeo WW (eds) Beyond description: Singapore space historicity. Routledge, London, pp 17–29

Yeung HWC, Savage V (1995) The Singaporean image of the orchardscape. In: Yeoh BSA, Kong L (eds) Portraits of places: history, community and identity in Singapore. Singapore, Times Editions

Yung SS, Chan KB (2003) Leisure, pleasure and consumption: ways of entertaining oneself. In: Chan KB, Tong CK (eds) Past times: a social history of Singapore. Singapore, Times Editions, pp 153–181

Zukin S (1996) Space and symbols in an age of decline. In: King A (ed) Re-presenting the city: ethnicity, capital, and culture in the 21st-century metropolis. New York University Press, New York, pp 43–59

Zukin S (1995) The culture of cities. Basil Blackwell, Oxford

Newspapers/Articles

Channel News Asia (2016) Uncertain future for Orchard road's pedestrian night. Singapore, 4 Feb 2016

CNN Travel (2012) Shopping in Singapore is better than Paris, 6 January 2012

Skyline (2001) Making Orchard road more happening! Singapore, URA, March/April 2001

Skyline (2002) Reclamation works at Pulau Ubin deferred. Singapore, URA, Jan/Feb 2002

Skyline (2003) Going public.Singapore, URA, Nov/Dec 2003

Skyline (2003) Ten ideas score for public space design. Singapore, URA, Nov/Dec 2003

The New Paper (2003) Orchard crush: don't go there? How can? 23 Nov 2003

The New Paper (2004) Foreign workers bowled over by cricket. 10 Feb 2004

The Straits Times (1996) The Singaporean: rich man, poor manners? 22 April 1996

The Straits Times (1997) 'Dead' stretch takes on new life. 16 Nov 1997

The Straits Times (2000) Eyes wide open. 12 April 2000

The Straits Times (2001) Moves to tackle Little India noise issue. 27 Oct 2001

The Straits Times (2003a) Disquieting silence: All's too quiet on the war front here. 22 March 2003

The Straits Times (2003b) Govt's Sars action swift, but shows up lack of checks. 10 May 2003

The Straits Times (2003c) The natives are restless. 30 Nov 2003

The Straits Times (2004a) The curse of being Singaporean. 22 Feb 2004

The Straits Times (2004b) Nursing mums make a stand at the Esplanade. 20 June 2004

The Straits Times (2004c) The nails-and-barbed-wire-mentality. 30 Sept 2004

The Straits Times (2004d) Welcome, not stifle, public activism. 20 Nov 2004

The Straits Times (2004e) Anti-casino groups keep up the fight. 26 Nov 2004

The Straits Times (2004f) Melting pot idea is idealistic, not realistic. 26 Nov 2004

The Straits Times (2004g) 3 wooden tables, a deck of cards and a wake. 6 Dec 2004

The Straits Times (2013a) Little India riots: police install 26 more CCTV cameras in Race Course, Buffalo roads. 11 Dec 2013

The Straits Times (2013b) Little India riot: take immediate action to protect residents. 12 Dec 2013

The Straits Times (2015a) Stricter rules in Geylang and Little India as alcohol laws kick in. 31 March 2015.

The Straits Times (2015b) GE2015: PAP vote share increases to 69.9 %, party wins 83 of 89 seats including WP-held Punggol East. 12 Sept 2015

The Straits Times (2015c) More pedestrian nights to come for Orchard road. 29 Dec 2015

The Straits Times (2016a) Singapore needs cultural shift to become a city for people, not for cars: Lawrence Wong. 30 Jan 2016

The Straits Times (2016b) Talking about Singapore's future: half-time report on the SGfuture dialogues. 21 Feb 2016

The Straits Times (2016c) Voiding the kampong spirit? 17 March 2016

The Sunday Times Forum (1986) Ah bengs sniff glue, we do not. 9 Feb 1986
The Sunday Times (2013) Lee Kuan Yew on fate of Singapore in 100 years' time. 11 Aug 2013
The Washington Post (2005) Wide 'Open' spaces here and abroad. 28 Jan 2005
Today (2015) Singapore's 'key future challenges': economy, population, identity. 30 June 2015
Urban Solutions (2004) Inequality and the city: the Singapore story. 4:96–107

Interviews/Blogs

Hughes G (1998) Range of vision: wanderings in the East/West world. http://glennh.tripod.com/ROVWhat.htm
Kiessling C (n.d.) Steps to insanity blog. http://www.angelfire.com/ga2/stepstoinsanity/Journal/page6.htm
Katakam S (2015) Interviewed by Subhas Nair in November 2015
Sa'at AB. Corridor. Blog entry in www.diaryland.com. 11 Sept 2004
Sa'at AB. Railings. Blog entry in www.diaryland.com. July 2004.

Official Speeches

Lee KY (1980) Speech by the Prime Minister, Mr Lee Kuan Yew, at the PAP 25th anniversary rally held at the National Theatre on 20 Jan 1980
Lee BY (1999) Speech by Dr. Lee Boon Yang Minister for Manpower and Member of Parliament for Jalan Besar GRC at the official launch of the sunday market at Weld road, Little India on Sunday, 3 October 1999, 7.00 pm at Jalan Besar neighbourhood police post
Mah BT (2001) Speech by Minister for National Development, Mr Mah Bow Tan, at the ground breaking ceremony for the first 40-storey HDB apartment blocks at Toa Payoh on Friday, 15 June 2001 at 10.30 am
Balakrishnan V (2002) Keynote speech by Dr Vivian Balakrishnan, Minister of State (National Development) For The IPS-NUS forum on "Civic Spaces in the Cities of the Asia Pacific," Institute of Policy Studies, 4 Mar 2002, 9.15 am
Lee HL (2003) Speech by Deputy Prime Minister Lee Hsien Loong at Pre-University Seminar 3 Jun 2003, Singapore @WAP
Prime Minister's Office Singapore (2014) Transcript of Prime Minister Lee Hsien Loong's speech at Smart Nation launch on 24 November. http://www.pmo.gov.sg/mediacentre/transcript-prime-minister-lee-hsien-loongs-speech-smart-nation-launch-24-november. Accessed 23 June 2016

MIX
Papier aus verantwortungsvollen Quellen
Paper from responsible sources
FSC® C105338

If you have any concerns about our products,
you can contact us on
ProductSafety@springernature.com

In case Publisher is established outside the EU,
the EU authorized representative is:
Springer Nature Customer Service Center GmbH
Europaplatz 3, 69115 Heidelberg, Germany

Printed by Libri Plureos GmbH
in Hamburg, Germany